A STRANGE WILDERNESS

A STRANGE WILDERNESS

The Lives of the Great Mathematicians

AMIR D. ACZEL

STERLING
New York

STERLING
New York

An Imprint of Sterling Publishing
387 Park Avenue South
New York, NY 10016

ISBN 978-1-4027-8584-9 (hardcover)
ISBN 978-1-4027-9085-0 (ebook)

Distributed in Canada by Sterling Publishing
c/o Canadian Manda Group, 165 Dufferin Street
Toronto, Ontario, Canada M6K 3H6
Distributed in the United Kingdom by GMC Distribution Services
Castle Place, 166 High Street, Lewes, East Sussex, England BN7 1XU
Distributed in Australia by Capricorn Link (Australia) Pty. Ltd.
P.O. Box 704, Windsor, NSW 2756, Australia

Book design by Level C
Please see photo credits on page 273 for image copyright information

For information about custom editions, special sales,
and premium and corporate purchases, please contact Sterling Special Sales
at 800-805-5489 or specialsales@sterlingpublishing.com.

Manufactured in the United States of America

2 4 6 8 10 9 7 5 3 1

www.sterlingpublishing.com

Frontispiece: The wilderness of the Pyrenees lies just beyond
the Aralar mountain range in northern Spain.

For Debra

Les fleuves lavent l'Histoire.

—J. M. G. Le Clézio

Mathematics is not a careful march down a well-cleared highway, but a journey into a strange wilderness, where the explorers often get lost.

—W. S. Anglin

$$a_{n+1} = \frac{a_n + b_n}{2} \qquad b_{n+1}$$

$$a_0 = 1 \qquad b_0 = \frac{\wedge}{1}$$

$$e^{i\varphi} = \cos\varphi + i\sin\varphi$$

Im

i

$\sin\varphi$

1 Re

φ

0 $\cos\varphi$

$$b_{n+1} = \sqrt{a_n b_n}$$

$$b_{n+1} = \sqrt{a_n b_n}$$

$$\frac{a_n + b_n}{2}$$

$$\frac{\pi}{4} = \sum_{n=1}^{\infty} \frac{1}{n}\left(\frac{3}{q^n - 1} - \frac{4}{q^{2n} - 1}\right.$$

$$a_{n+1} = \frac{a_n + b_n}{2}$$

$$a_0 = 1 \qquad b_0 = \frac{1}{\sqrt{2}} \qquad t_0 =$$

$$\sum_{n=}^{\infty} = \frac{\pi}{24}$$

CONTENTS

PREFACE

fell in love with the history of mathematics and the life stories of mathematicians when I took my first "pure math" course as a mathematics undergraduate at the University of California at Berkeley in the mid-1970s. My professor for a course in real analysis (the theoretical basis of calculus) was the noted French mathematician Michel Loève. A Polish Jew who by chance was born in Jaffa, in Turkish Palestine, he then moved to France, survived the dreaded concentration camp at Drancy (just outside Paris), and after the war immigrated to America. Loève was a walking encyclopedia of the rich intellectual life of mathematicians living in Paris in the period between the two world wars. He peppered his difficult lectures—which he delivered in abstract mathematical spaces, rarely deigning to "dirty our hands," as he put it, "in the real line," where all the applications were—with fascinating stories about the lives of famous mathematicians he had known and worked with. "We were all sitting at a café on the Boulevard Saint-Michel on the Left Bank, overlooking the beautiful Luxembourg Gardens, on a sunny day, when Paul Lévy brought up the mysterious conjecture by . . ." was how he would start a new topic.

So besides real analysis, Loève also taught us that mathematicians can live exciting lives, that they like to congregate in cafés—just as Sartre, de Beauvoir, and Hemingway did—and that they form an integral part of the general culture, or, rather, a fascinating subculture with its own peculiarities and idiosyncrasies. My interest was so piqued that later, also at Berkeley, I took a course dedicated to the history of mathematics, taught by the renowned logician Jack Silver. There I learned that the lives of mathematicians can at times be downright weird: they can get absurdly involved in grandiose political intrigue, become delusional, falsify documents, steal from each other, lead daring military strikes, carry on affairs, die in duels, and even perform the ultimate trick: disappearing completely off the face of the earth so that no one could ever find them. Silver himself was a bit of a strange mathematician: he dressed carelessly, was always disheveled, and when he finished what he wanted to say, he simply turned around and walked out of the classroom—never a "Good-bye," "See you next time," "I will be in my office from two to four," or any indication at all that class was over. We all sat there, looking at each other, until one intrepid soul or another would conclude that class was over and lead the way out. Presumably, Silver simply walked over to his office to continue working on his major theorem in the foundations of mathematics, Silver's theorem, which he proved that very same year.

As my mathematical career matured, I began to realize that behavior that may seem unusual elsewhere in society is often taken as "normal" in mathematical circles, where no one dares complain about it. My undergraduate adviser at Berkeley was the well-known topologist John Kelley. I loved Kelley so much as a professor that I ended up taking most of the courses he offered. But although he had a gift for making everything seem easier than it was, other things about his classes were hard to take— today, he would not be allowed to do many of the things he did in class. Kelley was never without a lit pipe in his mouth, and his syllabus read: "I smoke constantly in class, so if you worry about getting nicotine poisoning, don't take my course." He often brought into class his two huge dogs (apparently immune to nicotine), and they would plunk themselves down in front of him, blocking the aisle; when they scratched themselves, it was like a drumroll or a minor earthquake. Kelley's red or pink shirt was usually covered with political buttons, and he would admonish us to vote for his candidates or get involved in the political issues he favored.

After graduating, earning a doctorate some years later, and spending a dozen years teaching and doing research in mathematics and statistics as a professor, I returned to my early passion for the history of mathematics. And over the last decade and a half I have authored numerous popular books about the history of mathematics and the lives of mathematicians, from Fermat and Descartes to Cantor, Grothendieck, and the mysterious Bourbaki group. What I tried to do with all these books was to show how mathematics is entwined with the general culture, to point to what makes it unique and hence different from other disciplines, to expose how mathematicians think, and to showcase the kinds of exciting, interesting, and adventurous lives some mathematicians lead. I was greatly encouraged in this pursuit when Richard Bernstein, then a book critic at *The New York Times*, described in a book review the Parisian café scene presented in my 1996 book, *Fermat's Last Theorem*: "The scene . . . reminds us that the world has many worlds, with the priestly cult of mathematicians, so mystifying and inaccessible to most people, among the more esoterically interesting of them."

The realization of just how interesting the lives of mathematicians can be then caught the attention of a number of writers of both nonfiction and fiction. My friend Simon Winchester, author of superb works of nonfiction and acclaimed biographies, approached me at an international conference in Vancouver in 2001 and, in front of a crowd of three hundred people, suggested that we cowrite a biography of the nineteenth-century French genius Évariste Galois, who derived a remarkable theory in algebra and then died in a senseless duel at age twenty. The audience clapped enthusiastically as we shook hands in agreement. That book hasn't become a reality—at least not yet—but I am immensely grateful to Simon for encouraging my research into the life of Galois, which is discussed at length in one chapter in this present book.

Famous novelists, too, were attracted to the rich content and texture of the lives of major mathematicians. When my book *Descartes's Secret Notebook* appeared in Italian translation in 2006, Umberto Eco devoted his weekly column in the Italian news magazine *L'Espresso* to a thorough review of it and raised a number of interesting issues. After I responded in a letter, Eco invited me to visit him in Milan to discuss the life of Descartes. I will never forget the experience of standing with him in his thirty-thousand-volume library, which takes up much of the space of his apartment, browsing through original seventeenth-century manuscripts about the life of Descartes while sipping Calvados. I thank Umberto warmly for sharing with me his extensive knowledge of Descartes and his work, and for his enduring friendship. Descartes is the subject of one of the chapters in this book.

Three years after my book *The Mystery of the Aleph*, about the life of the tormented German mathematician Georg Cantor and his stunning discovery that there are various levels of infinity, was published in 2000, the novelist David Foster Wallace wrote his own account of the life of this great mathematician. Cantor often worked in a frenzied state and suffered frequent periods of depression. These dark moods, in fact, may not have been too different from Wallace's own bouts of depression, which reportedly may have been the cause of his tragic suicide. But in an interview for the *Boston Globe* in 2003, Wallace dismissed any connection between genius and madness and distanced himself from Cantor's mental problems, saying that he did not want to follow my approach of looking jointly at Cantor's mathematics and psychology; his book explored other directions, he said. But when Cantor died in 1918 in an asylum in Halle, Germany, he had been working on a mathematically impossible problem called the continuum hypothesis. His psychology, in fact, could not be separated from his mathematical work: late in life, Cantor made the continuum hypothesis a matter of personal dogma, "decreed by God." Some years ago, I visited the mental health facility in which Cantor spent years trying to recover, and where he ultimately died. A century later, the building is still a functioning hospital in an economically depressed part of Germany. I stood in the very room in which Cantor had worked on mathematics. I saw the claw-footed bathtub in which he was forced to immerse himself for hours as "treatment" for his depression, and I read the hospital's records of his repeated admissions and discharges from the late 1800s to 1918, when he died of starvation inside this facility. The life of Georg Cantor is covered in another chapter in this book, and I am grateful to the late David Foster Wallace for highlighting the question of the relation between psychology and our attempts to tackle the immense complexity of the infinite.

Sometime after my book *The Artist and the Mathematician*, about the lives of the mathematicians André Weil and Alexander Grothendieck, as well as the secret mathematical group called Bourbaki, was published in 2006, Sylvie Weil, André's daughter, published a poignant memoir about her life with her famous father and aunt. Her aunt was the philosopher Simone Weil, who often accompanied her brother to mathematical conferences of the Bourbaki group and became affectionately known as Bourbaki's "mother." Visiting Boston in 2011, Sylvie graciously shared with me many of the details I had not known about her father's life, and I am indebted to her for her kindness. André Weil is discussed in the last chapter of this book.

This chapter, fittingly entitled "The Strangest Wilderness," also deals with the life of Alexander Grothendieck—the mathematician who managed to completely disappear from our world, hiding somewhere in the forests or foothills of the high Pyrenees, which separate France from Spain. I am grateful to Pierre Cartier, a leading French mathematician and member of the Bourbaki group, for sharing with me his knowledge of the life of Grothendieck as well as fascinating stories about the founding and collective work of Bourbaki. Two mathematicians working in Paris, who insist on anonymity, also provided details about the life of Grothendieck.

Researching this book has been one of the greatest adventures of my life as an author. It took me to faraway corners of the world, from the island of Samos, where Pythagoras was born, to southern Italy, to Beijing and Delhi, and to countless locations in Europe—all in search of intricate details of the lives of our greatest mathematicians. I thank the mathematicians Marina Ville and Scott Petrack, my good friends, for their help and for discussing with me mathematics and the life stories of some of the mathematicians described in this book. I am indebted to Barry Mazur of Harvard, Akihiro Kanamori of Boston University, Goro Shimura of Princeton, Ken Ribet of Berkeley, and Saharon Shelah of Hebrew University for fascinating details of mathematics and its history.

My warm thanks go to my friend and agent, Albert Zuckerman, president of Writers House, for his encouragement, direction, and support throughout the process of writing this book. I am immensely grateful to my editors at Sterling, Michael Fragnito and Melanie Madden, who first suggested that I write a book about the lives of the great mathematicians. Melanie's superb editing of the manuscript in all its stages turned a rough draft into a complete book. She possessed the right vision of how to organize the complex material into parts, chapters, and sidebars, making the stories come alive and their heroes shine. Melanie wisely made me see what needed further explanation or expansion and what could well be omitted, and I am deeply indebted to her for her great insight and talent. I am also grateful to Barbara Clark for her excellent copyediting, and to Joseph Rutt of Level C for his beautiful design.

Finally, I thank my wife, Debra, and my daughter, Miriam, for their enthusiasm for this project and for their many helpful suggestions and ideas. I hope the reader will enjoy the life stories of the great men and women that are the subject of this book: history's greatest mathematicians.

INTRODUCTION

ur story begins around five thousand years ago in the civilizations of Egypt and Mesopotamia. These two ancient centers of human habitation are called potamic—from the Greek word for "river" (*potamos*)—because they developed in major river valleys: the great Nile in the case of Egypt, and the Tigris and Euphrates in the case of Mesopotamia. These valleys provided fertile ground for the development of agriculture—a key technological advance that had originated in the Jordan River Valley some eleven thousand years ago. The first mathematicians—people who performed rudimentary estimation work—were the "rope pullers" of the Nile Valley, assigned to demarcate the boundaries between the fields of various owners after the waters of the Nile receded every year following the annual inundation. Early geometrical ideas were developed in response to these problems. Pulling ropes in a flat terrain such as the Nile Valley led to the very first ideas in what is now called Euclidean geometry, named after Euclid of Alexandria, who lived much later. It is the geometry of straight lines that we study in school today.

Equally, astronomical observations of stars and planets carried out in Egypt and in Mesopotamia led to developments in mathematics. The Babylonians, Assyrians, and other inhabitants of the Fertile Crescent of Mesopotamia kept voluminous records of astronomical phenomena, such as solar eclipses, movements of planets, and locations of stars. The analyses of these data led to basic advances in calculation. Trade and finance, along with astronomy, brought about the use of a *sexigesimal* (base-60) number system in Mesopotamia. Though much more complicated than our current *decimal* (base-10) number system, we still see remnants of

The Greek philosopher Thales proposed the first known mathematical theorem in history during his visit to the Pyramids of Giza in Egypt in the seventh century BCE.

The Plimpton 322 cuneiform tablet, which contains fifteen Pythagorean triples, was discovered near Sankarah, Iraq. Scholars estimate it was written around 1800 BCE.

the Mesopotamian number system today in our clocks and geometrical and trigonometric calculations (e.g., an hour is composed of 60 minutes, a minute of 60 seconds, and a circle has 360 [6 x 60] degrees). The cumbersome base-60 number system of the Mesopotamians still enabled them to understand squares, square roots, and other concepts thousands of years ago.

Babylonian and Assyrian clay tablets contain an abundance of early mathematical discoveries. For example, the celebrated Plimpton 322 tablet has a list of 15 "Pythagorean triples"—sets of three squared numbers where two of them added equal the third (e.g., 9 + 16 = 25, which, as we know, is $3^2 + 4^2 = 5^2$). This tablet and others—including one displayed in the Louvre in Paris with a geometrical design that looks uncannily like a graphic depiction of what we call the Pythagorean theorem—have led many historians of mathematics to conclude that the ancient Mesopotamians may have developed the theorem we now attribute to Pythagoras, who lived at least twelve centuries later.

While the Babylonians and Mesopotamians impressed cuneiform signs into wet clay that they later baked in the sun, the Egyptians

The Ahmes Papyrus was written in Thebes during the Second Intermediate Period of Egypt (ca. 1650–1550 BCE). Its introductory paragraph says that the papyrus presents an "accurate reckoning for inquiring into things, and the knowledge of all things."

developed another technology for keeping records. By cutting reeds found along the Nile into strips, laying them side by side in two layers, and then mashing the layers into a sheet, they made papyrus, which they wrote on using a reed pen and carbon-based ink. The Ahmes Papyrus, named after the scribe who wrote it (it is also known as the Rhind Mathematical Papyrus, named after the Scottish collector who bought it from an Egyptian antiquarian in the nineteenth century), contains many examples of arithmetic, algebra, and geometry, including ways of solving equations and elementary mathematical problems arising from commerce and other areas of life.

Ancient India also had a thriving mathematical community that studied rudimentary computations relevant to everyday life and astronomy. The numerals we use today evolved from early Hindu numerals developed by Indian mathematicians. But the first mathematicians we know anything about—and the ones who made mathematics into an abstract, powerful science and art—were the ancient Greeks.

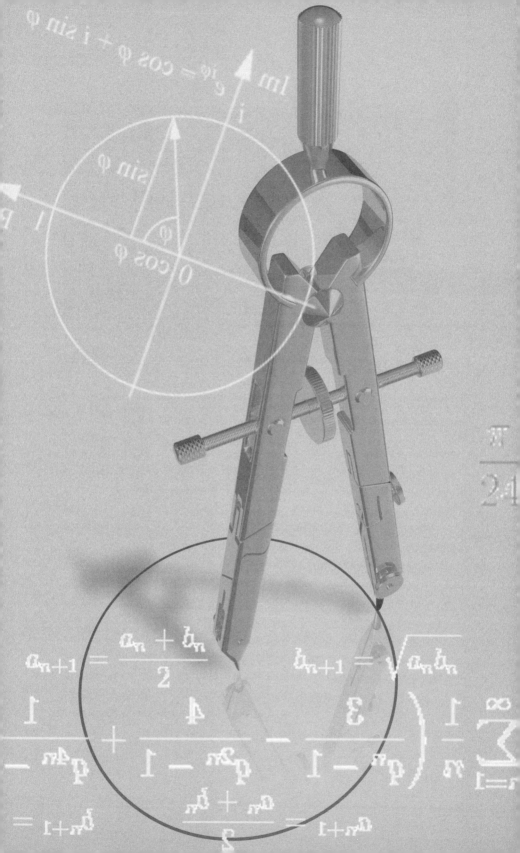

I

HELLENIC FOUNDATIONS

GOD IS NUMBER

T hrough the work of the Greeks, the early mathematics of the Babylonians and Egyptians changed
its character and became an abstract discipline
rather than a field mostly dedicated to the solution of practical
problems arising in astronomy or in everyday life. In creating
pure mathematics—mathematics divorced from any necessary
applications and constituting sheer knowledge—the Greeks
had achieved an intellectual advance of great power and
beauty.

THALES OF MILETUS

Mathematics as we know it today, with theorems and proofs, began with
the great Greek mathematician Thales of Miletus (ca. 624–548 BCE).
Miletus was among the first free city-states within the larger Greek

Miletus, in present-day Turkey, is the birthplace of the great Greek mathematician and philosopher
Thales. The ancient theater in this photograph was built during the fourth century BCE and expanded
during the Roman period.

empire, which spanned much of the eastern Mediterranean from Anatolia to the south of Italy and Egypt, including the islands in between. Lying on the coast of Anatolia, Miletus was one of the oldest and most prosperous Greek settlements of the time.

Thales is often called the first philosopher. He is also known for his famous saying "Know thyself," which was even engraved on the stone entrance to the cave of the Oracle of Delphi, a sacred site where the Greeks sought counsel from their gods. Additionally, Thales was one of the Seven Sages of Greece, though according to the historian Plutarch, he surpassed the others. In his book on Solon, another of the Seven Sages, Plutarch says this about Thales: "He was apparently the only one of these whose wisdom stepped, in speculation, beyond the limits of practical utility: the rest acquired the reputation of wisdom in politics."[1]

It is easy to forget just how ancient the Egyptian and Mesopotamian civilizations are, even as compared to Greece, which is to us an ancient civilization. By the time of Thales, who lived during the seventh century BCE, Egypt and Mesopotamia were already two millennia old. In the way a modern tourist may travel to Rome or Athens to view the magnificent ruins of the Roman Forum and Colosseum, or the Parthenon on the Acropolis, the Romans and Greeks who were contemporary with these monuments traveled to Egypt to view the pyramids and to absorb Egyptian culture. For example, the obelisks we see in Rome attest to how much the Romans loved Egyptian artifacts—so much so that they decided to bring some of them home. The obelisk at the Place de la Concorde in Paris attests to the greed of a far more recent visitor to Egypt: the French emperor Napoleon, who arrived on a voyage of conquest in 1798 (with two mathematicians, as we will later see) and plundered the land.

Like other young Greeks interested in philosophy and culture, Thales headed for Egypt, and when he arrived there, "he spent his time with the priests," as Plutarch tells us.[2] The priests taught him about Egyptian religion and philosophy, but he was also given the opportunity to practice some ingenious mathematics and subsequently to propose the first known theorem in history while visiting the pyramids.

Thales stood in the desert plain west of the Nile at Giza and looked up at the imposing Great Pyramid of Cheops, named after the Egyptian pharaoh Khufu (known as Cheops in Greece), for whose burial it was constructed. The colossal tomb was completed around 2560 BCE, so when Thales visited this huge edifice—still the most massive monument on earth and, over much of history, the tallest—the pyramid was two thousand years old. The Great Pyramid was one of the Seven Wonders of the Ancient World and is the only one still standing today.

Thales was awed by this pyramid, the largest and tallest—and, as we now know, also the oldest—of the three pyramids of Giza. Like any curious onlooker, he asked his Egyptian guides how high the pyramid was, but no one knew. When he asked the priests, they hadn't any idea, either. So Thales decided to measure the height of the pyramid without having to climb it—something that seemed impossible at the time. How could such a measurement be made from ground level?

The great pyramid is square-based, and each of its four sides lines up perfectly with the cardinal directions: north, south, east, and west. Hieronymus, a student of the Greek philosopher Aristotle, who lived two centuries after Thales, described what happened next. As quoted by the third-century-CE historian Diogenes Laertius: "Hieronymus says that he [Thales] even succeeded in measuring the pyramids by observation of the length of their shadow at the moment when our shadows are equal to our own height."[3] Thales knew how tall he was, so he waited for the moment in which his shadow was exactly the same length as his height, and he measured the length of the shadow cast by the Great Pyramid at that moment. The logic seems straightforward, but there are obstacles to take into account. Unlike a pole, or for that matter an obelisk of the kind Thales undoubtedly saw many times, a pyramid has bulk. Because it has a wide, square base, when the sun is high in the sky, the pyramid leaves no shadow along the ground but instead casts a shadow along its own slopes. However, being perfectly aligned with cardinal north, its shadow will exceed the extent of its base when the sun is low enough over the southern horizon. And when the sun is at the zenith—the highest point of its path through the sky—it will be perpendicular to the side of the base that faces north. But when will these conditions allow the shadow of an object to equal its height? This will happen when the sun's rays are at 45 degrees, twice a year—on November 21 and January 20 in Giza.

But did a great mathematician such as Thales have to wait for one of these two days? Not likely. A more probable scenario is described by the first-century historian Plutarch. In a fictionalized conversation between a Greek scholar named Niloxenus and Thales, the former says, referencing

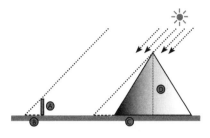

This diagram illustrates how Thales may have computed the height of the Great Pyramid using a stick of height *a*, which casts a shadow of length *b*. After computing the ratio *A/B*, he could then measure the length of the shadow plus half the length of the base of the pyramid (collectively, *c*). In order to find the height of the pyramid, *d*, all he has to do is multiply *c* by the ratio *A/B*.

the Pharaoh Amasis II, who ruled Egypt at the time: "Among other feats of yours, [the pharaoh] was particularly pleased with your measurement of the pyramid, when, without trouble or the assistance of any instrument, you merely set up a stick at the extremity of the shadow cast by the pyramid and, having thus made two triangles by the impact of the sun's rays, you showed that the pyramid has to the stick the same ratio which the shadow has to the shadow."[4] This calculation obviates the need to have the shadow equal the height of an object—i.e., the sun doesn't have to make an angle of 45 degrees—but the shadow must be longer than the half-length of the base. Half the length of the base must be added to the length of the pyramid's shadow, and a multiplicative ratio factor must be used. For example, if a yardstick's shadow is two yards long, then the length of the shadow of the pyramid (plus half the length of the base) must be halved to determine the height of the pyramid. When Thales implemented his method, he found a height of 280 Egyptian cubits, equivalent to 480.6 feet (though erosion has likely reduced the height of the pyramid slightly since that time). If Thales had used the former method on one of the two dates in which the angle of the sun's rays is 45

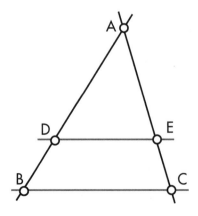

This illustration of Thales' intercept theorem contains a generalization of his pyramid-height calculation method, $d/c = a/b$. We have $DE/BC = AE/AC = AD/AB$.

degrees, however, then for a pyramid with a base length of 756 feet, half of which is 378 feet, the length of the shadow beyond the base would have been 102.6 feet.

But it is the ability to do the calculation on any day in which there is a shadow longer than half the base length by using a common proportionality factor that brought Thales to his beautiful theorem—the first theorem in history. This theorem, motivated by a real-world problem, is an abstract mathematical statement. In the figure at the bottom of page 6, two parallel lines intersected by two arbitrary lines cut segments according to the same proportion, illustrating Thales' theorem.

Thales' other theorems include the statement that the base angles of an isosceles triangle are equal and that a circle is bisected by a diameter. Until then, length, breadth, and volume were considered the key elements of geometry. Thales, however, was more concerned with beautiful geometrical theorems than the study of numbers, and he was the first mathematician to consider angles as important in the study of geometry. He provided a key link between triangles and circles by showing that every triangle corresponds to a *circumscribing* circle that touches all three points of the triangle. Thus he demonstrated that only one circle passes through any three points that are not all on the same straight line, and that the diameter of such a circle corresponds to the circumscribed triangle's hypotenuse. Additionally, he showed that an angle formed by the extension of two segments from the two endpoints of a diameter to any point on a semicircle is a right angle.

In addition to being the first "pure" mathematician, in the sense that he proposed and proved abstract theorems, Thales was also the first Greek astronomer. One day Thales was so engrossed in observing the

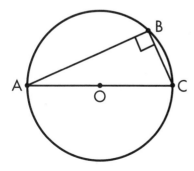

Thales' theorem shows that, if *A*, *B*, and *C* are points on a circle, where segment *AC* is its diameter, then *AC* also represents the hypotenuse of a right triangle.

stars that he moved forward a few steps without looking and fell into a well. "A clever and pretty maidservant from Thrace" who passed by and helped him out of the well chastised him for "being so eager to know what goes on in the heavens that he could not see what was straight in front of him, nay, at his very feet!"[5]

As an astronomer, Thales was so competent that he could predict solar eclipses. In fact, he is credited with predicting the total solar eclipse that took place in his part of Greece on May 28, 585 BCE. Greek mathematics historian Sir Thomas Heath explained that Thales' prediction was probably based on the fact that the Babylonians, who had been recording solar eclipses for centuries, knew that eclipses recur after a period of 223 new moons. Presumably, there had been a record of an eclipse in that area that had taken place 223 moons, or about eighteen years, earlier. This piece of information was probably transmitted to Thales through his intellectual connections in Egypt. He is also known to have studied the equinoxes and the solstices.[6]

PYTHAGORAS OF SAMOS

The next great Greek mathematician is the renowned Pythagoras of Samos (ca. 580–500 BCE). As a young man, he was coached by the aging Thales, and he would continue the Greek quest to turn the mathematics of the Egyptians, Babylonians, and early Indians from a practical computational discipline into a beautiful, abstract philosophy. It was Pythagoras who gave us the ubiquitous Pythagorean theorem, which allows us to determine the length of a right triangle's hypotenuse. Today GPS and maps use this theorem—as well as our very early understanding of numbers and of geometry—to compute distances between two locations.

Pythagoras was born on the Greek island of Samos, a stone's throw from the Anatolian Plateau of Asia Minor, which at that time was also part of greater Greece. The island is home to the Temple of Hera, one of the Seven Wonders of the Ancient World (although, unlike the almost-intact Great Pyramid, this temple has only one marble column still standing). Today the main town on the island is called Pythagoreion in honor of the island's native son.

Pythagoras began his life as a precocious intellectual adventurer,

The Greek philosopher and mathematician Pythagoras is depicted in this undated illustration.

curious about nature, life, philosophy, religion, and mathematics. As a young man, he traveled extensively. In Egypt he met with priests in temples to learn about their religion, their knowledge of the world, and their mathematics. In Mesopotamia he visited astronomers to learn how they observed celestial bodies, and he studied their mathematical and scientific methods. Did he learn about the theorem he is now credited with developing, or did he simply absorb related concepts in Mesopotamia? This we do not know. Because mathematics had roots in India as well, and because some Pythagorean ideas appear to be related to Indian mathematical principles, some historians have surmised that Pythagoras may have traveled as far as India. We have no confirmation of this conjecture, however.

Neither do we know how the great Thales met the young Pythagoras. We do know that the two men knew each other and that Thales recognized Pythagoras's budding intellect and encouraged him to expand his horizons. According to the third-century philosopher Iamblichus, who wrote a biography of Pythagoras, "Thales, admiring his remarkable ability, communicated to him all that he knew, but, pleading his own age and failing strength, advised him for his better instruction to go and study with the Egyptian priests."[7]

Pythagoras wanted to see much more than Egypt, so he first traveled east to Phoenicia, visiting Byblos, Tyre, and Sidon, where he met with priests and learned about Phoenician rites and customs. He is also reputed to have met with the descendants of the mysterious Mochus, a natural philosopher and prophet credited by some historians as proposing an atomic theory before Democritus. There Pythagoras is said to have been initiated into a strange regimen "to which he submitted, not out of religious enthusiasm, as you might think, but much more through love and desire for philosophic inquiry, and in order to secure that he should not overlook any fragment of knowledge."[8]

Pythagoras suspected that the rites and rituals he was observing and learning in Phoenicia had Egyptian roots, and he proceeded to Egypt to find their origin, just as Thales had encouraged him to do.

> There, he studied with the priests and prophets and instructed himself on every possible topic . . . and so he spent 22 years in the shrines throughout Egypt, pursuing astronomy and geometry and, of set purpose and not by fits and starts or casually, entering into all the rites of divine worship, until he was taken captive by Cambyses' force and carried off to Babylon, where again he consorted with the Magi, a willing pupil of willing masters. By them he was fully instructed in their solemn rites and religious worship, and in their midst he attained to the highest eminence in arithmetic, music, and the other branches of learning. After twelve years more thus spent he returned to Samos, being then about 56 years old.[9]

When he returned to his native island, Pythagoras was steeped in exotic ideas that he had absorbed during his travels. He developed a religious belief that the soul never dies but rather transmigrates to other living things. Hence, if a person kills another living thing—even a small insect—he could be killing a being with the soul of a deceased friend. This idea, which bears a strong resemblance to the Indian notion of reincarnation, led Pythagoras to a strictly vegetarian lifestyle. He also developed an aversion to eating beans—perhaps another fetish acquired as a result of his travels.

Pythagoras began to think about how he could combine the science of numbers and measurement that he absorbed in Egypt and Mesopotamia with the theorems of his Greek predecessor, Thales. Numbers fascinated him, so much so that eventually he and his followers would come to believe that "God is number." Further, Pythagoras transformed mathematics into the abstract philosophical discipline we see in pure mathematics today.

Pythagoras's notion that numbers held powers led to a kind of number mysticism, and he became a sort of guru. A growing group of disciples who adhered to his strict lifestyle principles and devoted their time to studying the abstract concepts of the new discipline of mathematics gathered around him. At some point a fearful island leader who worried that the group might someday vie for political power and unseat him applied political pressure on the Pythagoreans, and they were forced to leave Samos. Pythagoras and his followers moved to a place called Crotona, in the center of the bottom of the Italian "boot," which was also part of Magna Graecea (greater Greece). Isolated from the surrounding population, members of the secret society dedicated themselves to their religion—number mysticism—and the study of mathematics.

The Pythagoreans considered mathematics a moral beacon that helped them lead a righteous life. In addition to the word *philosophy* (love of wisdom), the word *mathematics,* which comes from the Greek phrase meaning "that which is learned," is believed to have been coined by Pythagoras.[10] He used both terms to describe the intellectual activity in which he and his followers were engaged. Pythagoras continued the work of Thales in pure mathematics and is seen to have transformed the discipline into "a liberal form of education, examining its principles from the beginning and probing the theorems in an immaterial and intellectual manner."[11] The "educational" aspect of mathematics was pursued in lectures that Pythagoras delivered to the members of his sect. These talks consisted of theorems, results, and discoveries about numbers and their meaning. As a form of public service to the outside community that surrounded the sect's compound—and, perhaps, to avoid being chased away, as had happened at Samos—Pythagoras gave public lectures to the entire community living in the area. The talks within the sect were strictly confidential, however. Most of the discoveries Pythagoras and his followers made about numbers were kept secret, with only select facts released to the outside world.

Mathematics historian Carl Boyer states, "The Pythagoreans played an important role—possibly the crucial role—in the history of

Pythagoras's triangular numbers.

mathematics."[12] What achievements by Pythagoras and his sect merit such an assessment? According to mathematics scholar Sir Thomas Heath, it was probably Pythagoras who discovered that the sum of successive natural numbers (i.e., 1, 2, 3, 4, 5 . . .) beginning with 1 makes a *triangular number*—that is, a number that can be drawn as a triangle. For example, 1 + 2 = 3; 1 + 2 + 3 = 6; 1 + 2 + 3 + 4 = 10; 1 + 2 + 3 + 4 + 5 = 15, and so on. Written algebraically, $T_n = 1 + 2 + 3 + 4 + \ldots + n = (\frac{1}{2})n(n + 1)$, where T_n is a triangular number and n is the number of units on a side.

DIVINE PROPERTIES OF NUMBERS

The second-century historian Lucian recounts how Pythagoras connected this property of the natural numbers with the Pythagoreans' number worship. One day he asked a member of his sect to count. The man began: 1, 2, 3 . . . When he reached 4, Pythagoras interrupted him and said, "Do you see? What you take for 4 is 10, a perfect triangle, and our oath."[13] Indeed, to the Pythagoreans 10 was a very special number.

Pythagoras and his followers saw the number 1 as the generator of all other numbers and the embodiment of reason. Two, the first even number, was considered female, representing opinion. Three was the first "true male" number, representing harmony because it incorporated both unity (1) and diversity (2). Four represented justice or retribution, since it was associated with the "squaring of accounts." Being the union of the first true male number (3) and the first female number (2), 5 represented marriage. The number 6 represented creation (and is the first "perfect number," as we will soon see), and 7 was the number of the Wandering Stars. (In addition to the sun and moon, the Pythagoreans knew of only five planets—Mars, Mercury, Jupiter, Venus, and Saturn—for which the days of the week are named. Sunday, Monday, and Saturday are obvious; for Tuesday through Friday, you can see the correspondence to the planets in their French forms: *Mardi, Mercredi, Jeudi,* and *Vendredi*.)

The number 10 was considered the holiest of holies—hence, Pythagoras's statement in the story above. It even had a special name, *tetractys,* from the Greek word for four (*tetra*), in reference to the number of dots to a side in the number's triangular form. Ten represented the universe as a whole, as well as the sum of the numbers that generate all the possible dimensions of the space we live in. (The number 1 generates all dimensions; 2 generates a line, since a line is created by the joining of two points; 3 generates a plane, since three points not all on a line determine a triangle—i.e., a two-dimensional figure—when joined together; four points, not all on a plane, generate a three-dimensional figure and, hence, three-dimensional space. And $1 + 2 + 3 + 4 = 10$.) Pythagoras and his followers called the number 10 their "greatest oath," as well as "the principle of health."[14] Of course, 10 is also the number of fingers and toes we have, from which fact our entire 10-based number system evolved and eventually superseded the base-60 system of the Babylonians and Assyrians. Equally, vestiges of a base-20 system (presumably emerging from the fact that, together, we have 20 fingers and toes) are still evident in the French language, where the word for 80 is *quatre-vingt* (four twenties).

Pythagoras was also interested in square numbers—like the triangular numbers, another set of "geometrical" numbers. As triangular numbers form triangles, square numbers can similarly be arranged to form squares. The first square number is 1 (by default, assuming it forms a square rather than a circle; indeed, $1^2 = 1$). The next square number is 4, then 9, then 16, and so on. If we draw these numbers as a two-dimensional figure, as the Pythagoreans did, we see the pattern in the figure on page 14.

To proceed from one square number to the next, we add the two sides of a square and add one. For example, to proceed from 4^2 to

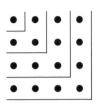

Pythagoras's square numbers.

5^2, we need to add $(2 \times 4) + 1$ to 4^2 (16). Indeed, 5^2 (25) is equal to 16 + $(2 \times 4) + 1$. Therefore, we can represent every square number as a sum of odd numbers: $(n + 1)^2 = 1 + 3 + 5 + \ldots + (2n + 1)$, where n is an integer.

In their search for mystical properties of numbers, the Pythagoreans defined a *perfect number* as a number that is "equal to [the sum of] its own parts." In other words, a perfect number is equal to the sum of all its multiplicative factors, excluding itself but including 1. The first perfect number is 6, because $6 = 6 \times 1$ and 2×3. As it happens: $6 = 1 + 2 + 3$. The next perfect number is 28, since $28 = 1 + 2 + 4 + 7 + 14$. The number 496 is also perfect. How do we know?

A few centuries later Euclid, the famous Greek mathematician, proved that if the sum of any number of terms of the series 1, 2, 2^2, $2^3 \ldots 2^{n-1}$ is a prime number, then that sum multiplied by 2^{n-1} is a perfect number. For example, for $n = 3$: $2^0 + 2^1 + 2^2 = 1 + 2 + 4 = 7$, a prime number. Therefore, 7 multiplied by $2^{(3-1)}$ must be a perfect number. It is, as $7 \times 4 = 28$. For $n = 4$, $1 + 2 + 4 + 8 = 15$, which is not prime. For $n = 5$, however, $1 + 2 + 4 + 8 + 16 = 31$, a prime number, so 31×16 should equal the next perfect number, 496.

The ancient Greeks also determined the following perfect number, 8,128, and began to notice a pattern: a perfect number always ends either in a 6 or in an 8. But further perfect numbers were beyond their computational ability—the fifth perfect number, 33,350,336, is very large, and the next one, 8,589,869,056, is in the billions. The ninth perfect number has 37 digits!

Pythagoras was once asked by a disciple, "What is a friend?" He replied, "A friend is an alter ego." This led him to define the concept of friendship for numbers as well, defining two numbers as being "friends" if each one is the sum of the multiplicative factor of the other number. Hence, the numbers 284 and 220 are friends. Why? $284 = 1 + 2 + 4 + 5 + 10 + 11 + 20 + 22 + 44 + 55 + 110$, which are all of the factors of 220, while 220 = 1 + 2 + 4 + 71 + 142, which are all of the factors of 284.

––––––––––––––––––––◆––––––––––––––––––––

Pythagoras and his followers also understood fractions, such as $2/7$ or $31/77$. We call such numbers *rational numbers,* perhaps because they make sense to us. A pie can be divided into seven pieces, each being $1/7$ of the whole, and you can give someone two pieces—a fraction of the entire pie represented by the number $2/7$. But when the Pythagoreans went further in their mathematical and mystical exploration of numbers and their properties, they ran into a conundrum that stunned them and perhaps even brought on their demise. This paradox—which came about in the interface between geometry and arithmetic—would come to a head with the work of Georg Cantor in the nineteenth century, and it continues to haunt us even today.

PLATO'S ACADEMY

Despite having so little foundational work to build on, the mathematicians of ancient Greece were able to make advances in mathematics that stun us today. Thanks to the pioneering work of Thales and Pythagoras, abstract mathematics became what we know it to be: a science—and, in many ways, an art—based on pure logic and consisting of results we call theorems (and lemmas and corollaries that precede and follow them, respectively), which must be proved. Many of their discoveries were so fundamental that they seem "modern" in their depth of meaning and influence, and as mathematics became interlinked with philosophy, it attracted the attention of the greatest philosopher of the time: Plato.

Raphael's famous fresco *The School of Athens* (1510–11) echoes the notion of Plato's Academy as an intellectual community of learned scholars. Among the thinkers depicted in the painting is Plato himself, who stands at the center of the image, pointing upward.

HIPPASUS OF METAPONTUM

Plato's dialogues show that he and the members of his Academy in Athens had been stunned by a discovery about the nature of numbers that made their philosophical view that "God is number"—where "number" means a whole number or a fraction made of two integers—shaky. The discovery was made by members of the Pythagorean order sometime before 410 BCE. Some early historians attribute this shocking discovery to the Pythagorean mathematician Hippasus of Metapontum (another name for Crotona, the Pythagoreans' centuries-long abode). Hippasus was reportedly expelled from the order—or worse—because of what happened next.[1]

In addition to the study of numbers, the Pythagoreans did much work on pure geometry, as initiated by the work of the great Thales and pursued very actively by Pythagoras himself. Hoping to associate numbers with lengths in geometry, an objective that seemed simple enough, Hippasus drew a square in the sand. As soon as a diagonal was drawn inside the square, however, the question arose: If the length of the side of the square is one unit, what is the length of the diagonal? The result of this elementary investigation would, in fact, change the world of mathematics.

Today nearly every schoolchild knows the famous Pythagorean theorem, which states that the sums of the squares of the two sides of a triangle that share a right angle between them is equal to the square of the hypotenuse.

So when Hippasus drew a square in the sand and added the diagonal, it was clear to all witnesses that the diagonal was the hypotenuse of a right triangle, with opposite sides equal to one unit each. Based on the Pythagorean theorem, the length of the hypotenuse must be the square root of 2.

But what is the square root of 2? We know that the square root of a square number is an integer (e.g., the square root of 4 is 2, the square root of 9 is 3, and so on). Two is not a square number—so what is its square root? Until that point, the Pythagoreans knew of only (positive) integers and what we call *rational numbers*: fractions made of two integers—a numerator and a denominator. The square root of 2 is clearly not an integer, since 2 is not a square number, so the Pythagoreans assumed that it was a rational number. They searched in vain for the two integers that made up the numerator and denominator of the fraction but came up empty-handed. It can be proved mathematically that such integers do not exist.[2]

It is believed that Hippasus broke the Pythagoreans' code of secrecy by revealing to the outside world the existence of numbers that could not be written as fractions of integers. One story has him simply expelled from the order for this crime, but according to another story, the Pythagoreans erected a tombstone with his name on it and presumably killed him. According to yet another version, he was forced on a sea voyage from which he did not return. All this for revealing the existence of *irrational numbers*, which comprise many other square roots, higher-order roots (e.g., the cube root of 2), the natural numbers π and e, and infinitely many others. We will learn about these numbers in detail when we arrive at the nineteenth century.

ANAXAGORAS OF CLAZOMENAE

Anaxagoras of Clazomenae lived in the fifth century BCE. At the beginning of this century, the Persian invaders of Greece were defeated, and at the end of the century, Athens was defeated by Sparta. The period between these two key events is called the Age of Pericles. Known for its great achievements in art and literature—including the amazing statues and plays that form much of the foundation of Western composition—it was an age of peace and prosperity, as well as flourishing intellectual activity.

Athens attracted the greatest mathematicians of its vast dominion. Zeno of Elea (in southern Italy) gave us the famous paradoxes about time, space, and infinity, including the notable one about Achilles and the tortoise, in which Achilles races against the tortoise but cannot win because each time he covers half the distance to the tortoise, the tortoise has already advanced farther, and when Achilles has covered half that distance, the tortoise has advanced again, and so on. This period was also the era of Democritus, who proposed that the universe is made of atoms—something that would be proved by scientists almost two and a half millennia later.

A deep thinker who concerned himself with uncovering the structure of the cosmos, Anaxagoras could well be called one of the earliest natural philosophers in history. His birth date is unknown, but we know that he died in the year 428 BCE and that he came from Clazomenae in Ionia, a

Anaxagoras, perhaps the first mathematician to tackle one of the three classical problems of antiquity, is portrayed in this detail from the 1888 fresco *Philosophers of Athens*.

region of western Greece that includes the islands of Corfu and Ithaca. After acquiring a degree of fame for his novel views about nature and the universe, Anaxagoras became a tutor to the great leader Pericles, but his ideas were so ahead of his time that he was frequently imprisoned for heresy.

Anaxagoras shocked many Athenians by publicizing his view that the sun was not a god but rather a flaming rock in the sky that was as large as the entire Peloponnesian peninsula. He contended that the moon was another planet, like Earth, and that it had its own population and reflected light from the sun. Many people took offense at his unprecedented theories, and accusations of heresy led to his imprisonment. Anaxagoras's friend and protector Pericles eventually forced his release from prison, but the more he spouted his ideas, the longer he found himself behind bars.

During his imprisonment, Anaxagoras used to while away the time by working on a mathematical problem that no one could solve. According to Plutarch, he was trying to square the circle. Plutarch's account is the first mention in history of a problem that would consume the time and effort of many a mathematician until proved impossible in the nineteenth century. What is so beautiful about this problem is that it is a purely intellectual one—as was most of Greek mathematics.

THE THREE CLASSICAL PROBLEMS OF ANTIQUITY

Squaring the circle is the first of the celebrated "three classical problems of antiquity." All three problems are now known to be impossible, thanks to the work of Évariste Galois (1811-32), whose genius and tragic life are discussed later in this book. The problem of squaring the circle presented the challenge of constructing a square with the same area as the area of a given circle. Later sources than Plutarch tell us that only a straightedge and a compass—the two tools that Greek construction employed—were allowed to be used in this effort. Although Anaxagoras is best known for his work in natural philosophy, he is credited with the first known attempt to solve this problem.

The second of the three famous problems is called doubling the cube. Also known as the Delian problem, the story of its genesis is both stirring and sinister. In 429 BCE a great plague raged in Athens, killing a quarter of the city's population. Persistent and recurring, the scourge is believed to have claimed the life of Pericles, and as we will see, it formed the basis from which the famous second classical problem of antiquity emerged.

The island of Delos lies in the Aegean Sea, a body of water in the eastern Mediterranean that stretches from mainland Greece down to Crete. A small, now arid island, it is the navel of the Cyclades, a chain of islands that appears to form a "cycle" around it. Because of its special position in the center of the islands, the ancient Greeks considered Delos to be a holy place, and each of the Greek city-states built temples dedicated to their gods on the tiny landmass. Ruins of the many temples built on this island can still be seen today. (The island is easily accessible by boat from nearby Mykonos.)

When the plague first began to spread in Athens around 430 BCE, the Athenians scrambled to try to save themselves from what they viewed as the wrath of the gods. They sent a delegation to Delphi, located on the Greek mainland not far from Athens, to ask the oracle to intercede with the gods on their behalf. The oracle came back with the pronouncement: "Apollo wants you to double his temple on Delos."

The Athenians frantically set to work. They doubled the length, width, and height of the Athenian temple to Apollo on Delos. But the plague continued to rage. So again the Athenians sent a delegation to the Delphic oracle to find out why Apollo was still angry with them and to plead for an end to the plague—after all, they had done exactly what he had asked them to do. "No," replied the oracle. "You haven't done as the god had instructed you to do. Go back to Delos and do as he has commanded you!"

The Athenian masons and engineers soon realized why they had failed: Apollo wanted the *volume* of his cubic temple doubled—not all its dimensions. By doubling each side of this cube-shaped temple, what they had inadvertently done was to increase the volume eight-fold ($2 \times 2 \times 2 = 2^3 = 8$). What they needed to do in order to double the volume of the temple while maintaining its cubic dimensions was to increase the length, width, and height of the cube by a factor equal to the *third root of 2*. Only this way could a cubic volume be doubled, since $(2^{1/3})^3 = 2$. The engineers, masons, and builders realized that they needed to go back to the original cubic temple and, using the only tools of their trade—a straightedge and a compass—expand each of the three dimensions so that its length would increase by a factor of exactly the cube root of 2. But it couldn't be done, and the plague raged on. As we will see, the work of Évariste Galois in the nineteenth century also proved that the Delian problem inspired by this story—the problem of doubling the volume of a given cube using only a straightedge and a compass; i.e., to construct, geometrically, a length that is the cube root of 2 times as large as a given length—is impossible.

The third of the so-called classical problems of antiquity is a problem that, like the two discussed above, was circulating in intellectual circles in Athens at that time. It is the problem of trisecting an arbitrary angle using only a straightedge and a compass. With these tools, some angles can be trisected, meaning that given an angle, one can sometimes construct an angle that is a third as large. But the problem is to be able to do this trisection with any given angle. Archimedes, discussed later in this chapter, was able to use the famous Archimedean spiral to trisect angles, but his method required more than just a straightedge and compass, so he failed to solve the original problem.

PLATO

The Greek philosopher Plato (428–348 BCE) was born just a year after the infamous plague. He was not a mathematician, but he believed that mathematics was the study of truth. To emphasize this view, Plato placed a sign over the gate of his Academy in Athens: LET NO ONE IGNORANT OF GEOMETRY ENTER HERE! Plato thus became known as the maker of mathematicians, and he encouraged many geometers and algebraists to join what would later be considered the first center of philosophy and knowledge in the world.

Despite Hippasus's discovery that not all numbers were expressible as ratios of natural numbers, Plato was undeterred in his love of numbers and in his belief in their divinity, and he remained faithful to the Pythagorean ideal of number until death.[3]

Mathematicians of ancient Greece knew that there were five, and only five, *regular solids*. These are three-dimensional geometrical objects whose faces are all identical to one another. The solids that satisfy this requirement are the cube, the tetrahedron (triangle-based pyramid), the octahedron, the icosahedron, and the dodecahedron. Plato admired the discovery and ascribed much importance to these solids; so they became known as the Platonic solids. The Greeks associated the five solids with elements of nature: earth, water, air, and fire, as in the figure on page 24.

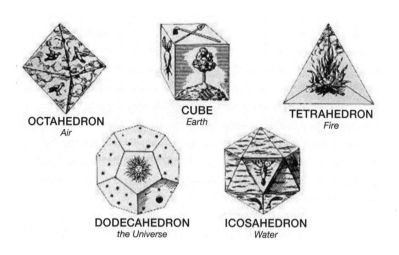

OCTAHEDRON
Air

CUBE
Earth

TETRAHEDRON
Fire

DODECAHEDRON
the Universe

ICOSAHEDRON
Water

This illustration of the five Platonic solids appeared in Johannes Kepler's *Mysterium Cosmographicum*, published in 1596.

Of the five solids, Plato considered the dodecahedron so important that he named it the "fifth essence," or quintessence, from which we get the word *quintessential*.

EUDOXUS OF KNIDUS

Eudoxus was born in 408 BCE to an impoverished family in Knidus, Asia Minor. Because of his family's low socioeconomic status, he would have had no chance at a successful life if it weren't for his powerful mathematical skills. As a young adult, Eudoxus heard about Plato's Academy and borrowed money to travel there. Many of the philosophers at the Academy ignored the young man, but Plato recognized his genius and supported him in his mathematical pursuits.

There was no remuneration for membership in the esteemed Academy, and Eudoxus had so little money that he could not afford to live with the other members in Athens. He was forced to rent a small room in the nearby city of Piraeus, where rents were low and basic food could be obtained inexpensively. He commuted daily to Athens to attend the discussions at the Academy. Eventually, after having proved several major theorems in geometry that no one had been able to tackle, Eudoxus earned the respect of the other philosophers. Thanks in part to the constant encouragement he received from Plato, Exodus surpassed all the mathematicians who came before him by devising the basic ideas of integral calculus 2,100 years prior to its formal and complete introduction by Newton and Leibniz.

Eudoxus was able to calculate volumes and areas using essentially the calculus ideas we use today. In fact, in modern mathematical analysis we use "Eudoxus sums" as part of the derivation of the integral. Unfortunately, the resulting envy of lesser mathematicians in the academy finally drove Eudoxus to leave Athens and settle in Cyzicus, where he learned and then practiced medicine. Exodus became very wealthy and was even elected a legislator.

ARCHIMEDES OF SYRACUSE

Archimedes (ca. 287–212 BCE) was a relative of the ruler of Syracuse, Hieron II, and his father was the astronomer Phidias. Belonging to one of Syracuse's most aristocratic families, Archimedes didn't have to work, and mathematics became his passion. In fact, he is believed to have cared so little about daily life that he left meals uneaten when a mathematical problem occupied his mind.[4]

Archimedes is famous for a great discovery he made while taking a bath. The episode with the bath actually begins with a request by Hieron to his bright relative to help him determine whether a goldsmith had stolen some of his gold. The king had apparently asked the goldsmith to make him a new crown, providing him with the needed gold. When the king got his new crown from the goldsmith, he became suspicious that the goldsmith had replaced some of the gold inside the crown with silver or another cheaper metal and then pocketed the missing gold. The king needed Archimedes to find a way to compare the density of the crown with the density of gold without damaging the crown.

Archimedes pondered this dilemma while taking a bath, absently noting that the tub water rose when he got in. Suddenly it hit him: the water displaced was equal in volume to that of his body. Famously, he jumped out of the bath and ran naked through the streets of Syracuse, shouting, "Eureka, eureka!" (I found it, I found it!)

The Greek mathematician Archimedes discovered that a sphere circumscribed within a cylinder of the same radius has a volume and area equal to two-thirds of the volume and area of the circumscribing cylinder.

Archimedes now had a method for measuring the volume of an irregularly shaped object, such as a crown or a human body, by immersing it in water and measuring the volume of the water it replaces. Then, by weighing the object—the crown, in this case—one could find its density (weight divided by volume). And by comparing this density with that of gold, one could determine whether the goldsmith had cheated the king and stolen his gold! We don't know whether the goldsmith was found guilty or not, however.[5]

Beyond this discovery, Archimedes helped his king and the people of Syracuse fight against the invading Romans by inventing many machines useful in warfare, including various kinds of catapults and (supposedly) a set of mirrors that focused the sun's rays on attacking ships, causing them to burn, although there's no strong proof that this last invention was viable. Many of the weapons he invented enabled the Syracusans to defend themselves—at least for a while—against Roman attack.

Archimedes also applied Eudoxus's amazingly powerful analysis to the study of the areas and volumes of solids. As such, his work also anticipated the development of calculus. He is also known for discovering the Archimedean spiral, which is the locus of all points traced through time as a point moves away from the origin at constant speed and as the coordinate system rotates at a constant angular speed. Another result, appealing in its simplicity, was Archimedes' discovery that a sphere circumscribed within a cylinder of the same radius has a volume and area equal to ⅔ of the volume and area of the circumscribing cylinder. Archimedes was so pleased with this result that he requested that a bronze sphere inside a cylinder be placed on his grave.

According to historical record, Archimedes' life was cut short while he was trying to prove a geometrical theorem by drawing in the sand. When

Despite orders from the Roman general Marcellus that Archimedes' life should be spared, the great scientist was killed by a Roman soldier during the Siege of Syracuse around 212 BCE. Syracuse had been protected by weapons designed by Archimedes.

a Roman soldier approached him, Archimedes cried, "Don't disturb my circles!" With that, he was promptly killed by a swing of the sword.

Many centuries later Europe revived the work of Eudoxus and Archimedes, culminating with the development of calculus. In the meantime, however, the mathematical treasures of the Greek world traveled to Arabia through Euclid's famous books, the *Elements*, and Europe descended into the darkness of the Middle Ages.

THREE

———— • ◆ • ————

ALEXANDRIA

reece declined as Rome gained influence and forced the infusion of the harsher and more pragmatic Roman lifestyle, which valued engineering, military goals, discipline, killing animals for entertainment, and law over the Greek values of democracy, philosophy, art, and theater. Yet the Greek art of mathematics continued to be pursued by Greek mathematicians, most notably in Egypt. Alexandria, which boasted the largest and most famous library of the time until fire destroyed it in 48 BCE, had a thriving mathematical community. In fact, Alexandria, which had been founded in 331 BCE by Alexander the Great, became the undisputed center for mathematics in the Western world and would remain so until the early Middle Ages.

Alexandria, Egypt, was home to many of history's most influential and important mathematicians, including Apollonius, Eratosthenes, and Diophantus. This undated illustration depicts scholars poring over scrolls in the ancient city's famed library.

This fragment from Euclid's *Elements*—one of the Oxyrhynchus Papyri, discovered in Egypt in 1896–97—contains a diagram illustrating proposition 5 of Book II. It is one of the oldest and most complete extant diagrams from Euclid's seminal work.

In the third century BCE, Alexandria had been home to Euclid, whose *Elements* include in their thirteen volumes all the great theorems of Thales, Pythagoras, and other Greek mathematicians, theorems that laid the foundation for geometry based on straight lines. Euclid also put forward five postulates summarizing the geometry of the ancient Greeks:

1. A straight-line segment can be drawn joining any two points.

2. Any straight-line segment can be extended indefinitely in a straight line.

3. Given any straight-line segment, a circle can be drawn having the segment as radius and one endpoint as center.

4. All right angles are congruent.

5. If a straight line falling on two straight lines makes the interior angles on the same side less than two right angles, the two straight lines, if produced indefinitely, meet on that side on which the angles are less than the two right angles.

Greek mathematicians took these postulates for granted as obvious starting points for the study of geometry. No one would ever argue with the first four postulates, but the fifth was very troublesome . . . and would be debated for millennia.

APOLLONIUS OF PERGA

Apollonius of Perga, Asia Minor, lived roughly between 260 and 190 BCE. Although not born in Alexandria, he spent considerable time there studying and working. Just thirty years younger than Archimedes, he pursued the kind of mathematics that had been developed by the celebrated mathematician and came up with what is famously known as the Problem of Apollonius: given three *things*—where "thing" may be a point, a line, or a circle—find a circle that is tangent to (i.e., touching but not intersecting) each of the three things. For points, the problem is easily solved as the circumscribing circle. For three circles, however, the problem is difficult. Almost two millennia later Newton finally came up with a solution using only a straightedge and compass, as had been required for the three Classical Problems of Antiquity.

Apollonius studied conic sections—geometric objects formed by making various cuts through a cone—and he named them the hyperbola, the parabola, and the ellipse. He also deduced their mathematical properties. These geometrical figures would hold the attention of mathematicians for millennia, reappearing in the work of Descartes, Fermat,

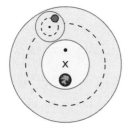

Ptolemy's model of the solar system, showing a planet rotating on an epicycle, which itself is rotating on a circular orbit (deferent) around X—a point halfway between Earth and the equant.

Newton, and Leibniz in the seventeenth and eighteenth centuries, leading up to the invention of calculus. The intense interest in conic sections can be attributed to the fact that they are not just cuts made through a cone; they are also two-dimensional figures that, as Descartes and Fermat would later show, can be described by purely algebraic methods. The simplest parabola, for example, can be described as the curve written algebraically as $y = x^2$, where x is measured along the horizontal axis and y along the vertical. In making this connection between conic sections and algebra, Descartes would launch a powerful link between algebra and geometry.

Apollonius also invented the concept of epicycles—circles centered on larger circles—which are mathematically valid constructions but which received a bad reputation from our modern vantage point because of their famous misapplication in astronomy by Ptolemy of Egypt in his second-century treatise the *Almagest* (Great Compilation). Ptolemy's model of the solar system placed Earth at the center of creation and represented the motion of the planets as epicycles on a circular orbit centered on a point halfway between Earth and the *equant*. Ptolemy's model persistently stood in the way of progress in cosmology, giving fodder to the Catholic Church in its opposition to science and the Copernican system for centuries.

ERATOSTHENES OF CYRENE

Eratosthenes of Cyrene (ca. 276–194 BCE) was a young contemporary of Archimedes. While living in Athens, where he wrote poetry, studied history and mathematics, engaged in athletics, and observed the stars and planets, he showed great promise in many areas of art and science. His fame reached the ear of the ruler of Alexandria, Ptolemy III, who invited Eratosthenes to his city and offered him a position as librarian and

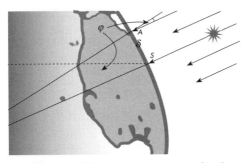

In this illustration of Eratosthenes' measurement of Earth's circumference, the sun appears directly over Syene (*S*) on the summer solstice. At the same time, the angle of the sun to a perpendicular to the surface of the earth (*φ*) is approximately 7°12', or ¹⁄₅₀ of a circle, at Alexandria (*A*). Multiplying the distance between Syene and Alexandria (*δ*)—roughly 500 miles—by 50 gives us an estimate of Earth's circumference: 25,000 miles.

tutor to his son. While Eratosthenes was in Alexandria, Archimedes sent him his treatise *The Method of Mechanical Theorems.* As the young mathematician studied the work and learned more about astronomy, he became interested in what appeared to be an extremely difficult problem: measuring Earth's circumference.

Perhaps aware of the work of the "first mathematician," Thales, Eratosthenes thought up an ingenious method for measuring Earth's circumference using the sun and shadows. Eratosthenes found that on the day of the summer solstice, June 21, the sun shone directly down into a deep well at Syene (present-day Aswan) in Upper Egypt. This meant to him that the sun was directly overhead—it was perpendicular to Earth's surface at that point. At the same time, an associate at Alexandria, which Eratosthenes took to be roughly on the same meridian (it is slightly west, in reality), measured the angle of the sun to a perpendicular to Earth's surface *5,000 stades*—roughly five hundred miles—to the north. At Alexandria the angle of the sun to the perpendicular was measured as ¹⁄₅₀ of a circle.

As seen in the picture, the angle of ¹⁄₅₀ of a circle is identical to the angle measured from the center of Earth. Eratosthenes deduced that since 5,000 stades equal ¹⁄₅₀ of Earth's circumference, the full circumference must be 50 × 5,000 stades, or 250,000 stades (about 25,000 miles—very close to our modern value for the circumference of Earth!).

Eratosthenes' calculation was undoubtedly one of the greatest achievements of the time. But Eratosthenes did much more. Among his achievements in pure mathematics is an advance in number theory. He invented the Sieve of Eratosthenes, a device that enables us to determine prime numbers less than or equal to any given number (although it works well only for relatively small numbers, because the computation becomes tedious and inefficient for large sets). To see how it works, let's use an example: find all the prime numbers between 2 and 100.

Begin by listing all the numbers between 2 and 100, as shown below:

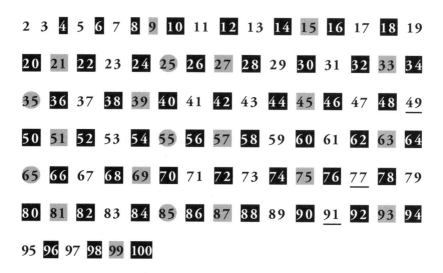

Then cross out all multiples of 2 (all even numbers, here appearing against a black background), after having identified 2 as the first prime number (1 is not considered a prime number for various technical reasons). Next, proceed to the next prime number, 3, mark it as prime, and cross out all further multiples of 3 that have not been already eliminated (here shown against a gray background). Then proceed to 5, mark it as the next prime number, and cross out all remaining multiples of it (here shown within a gray circle) until you reach the end of the list. The process continues with 7 (multiples of which are underlined here), and when we reach 11 there are no new numbers to cross out, so the remaining set of numbers represents all prime numbers between 2 and 100. Today, with computers and better technology, we can go quite far in our search for prime numbers, but Eratosthenes' systematic search laid the groundwork and constitutes the first-ever algorithm for finding prime numbers.

DIOPHANTUS

With the influence of the Roman Empire, Greek mathematics of later centuries became increasingly practical, and so-called pure mathematics—mathematics regarded simply as an intellectual exercise and a philosophy—declined. Among the great Greek mathematicians of Alexandria during the first few centuries CE were Pappus (ca. 290–350), who achieved important strides in geometry; Proclus (410–85), who wrote an influential commentary on Euclid's *Elements*; Diophantus, the Father of Algebra, who lived in the third century CE; Theon (ca. 335–405), who published an authoritative edition of Euclid's aforementioned monumental work; and his daughter Hypatia (d. 415).

Diophantus of Alexandria made key advances in algebra, and Diophantine equations—for example, $x^2 + 2y^3 = 25$—are studied in number theory today. These equations have several variables appearing in various powers. Diophantus was the first mathematician in history to study in detail such equations and to recognize that rational numbers (ratios of integers) can be used as both coefficients and solutions.

All we know about Diophantus is that he probably lived around the year 250 CE. A mathematical riddle that appeared in a book of puzzles by an unknown author, written sometime in the fifth or sixth century, offers a few clues about his life:

> God granted him boyhood for the sixth part of his life, and adding a twelfth part to this, He clothed his cheeks with down; He lit him the light of wedlock after a seventh part, and five years after his marriage He granted him a son. Alas! Late-born wretched child; after attaining the measure of half his father's life, chill Fate took him. After consoling his grief by this science of numbers for four years, he ended his life.[1]

Diophantus's mathematics was far more complicated than this simple linear equation implies. Solving this riddle, the solution to which is 84, tells us a few things about Diophantus, such as the age he was when he married, his age when his son was born, the age he was when he died (eighty-four), and so on.

Diophantus wrote his mathematical results in a series of books called the *Arithmetica*, only six of which survive. His work was largely lost to history until Pierre de Fermat consulted a Latin translation of the *Arithmetica*, and in its margin he wrote his famous Last Theorem—a generalization of a result by Diophantus that was proved by Andrew Wiles of Princeton University in the 1990s. The centuries-old interest in solving Fermat's problem refocused the attention of mathematicians from the seventeenth century onward on Diophantine equations—i.e., the analysis and solution of algebraic equations with several variables and the attendant study of polynomials (sums of multiples of powers of a particular variable, or of several variables)—and brought Diophantus fame as the Father of Algebra.

HYPATIA

The daughter and pupil of Theon, a notable scholar and mathematician of the era, Hypatia (ca. 360–415) was the first important female mathematician in history. Fittingly, her name is derived from the Greek word for "highest." Hypatia studied with her father and at an early age became interested in philosophy and mathematics.

Hypatia pursued mathematics and philosophy and became a lecturer in philosophy at the Neoplatonist school in Alexandria. Many of her students came from far away to hear her lectures. She was, in a way, a philosophical descendant of Plato, continuing his academy's tradition of lecture and discussion. In addition to teaching the philosophy of Plato and mathematics, Hypatia wrote commentaries on Diophantus's *Arithmetica,* Apollonius's *Conics,* Euclid's *Elements,* and Ptolemy's *Almagest.* She also wrote a book on astronomy, entitled *The Astronomical Canon.*

But Hypatia lived in turbulent political times that saw severe friction between paganism and the emerging religion of Christianity. As a woman who had achieved cultural and political influence in Alexandria, and as a freethinking intellectual who pursued a career in a world dominated by men and never married, she was vulnerable. Indeed, in March of 415 Hypatia, an alleged pagan, was publicly accused of using her political influence to prevent a reconciliation between the estranged Imperial Prefect at Alexandria and the Christian Patriarch.

A group of monks reportedly ambushed her chariot and dragged her to the street, stripped her naked, and killed her. Some versions of the story have her flayed alive, her limbs torn and burned. To some historians

Charles William Mitchell's 1885 painting of Hypatia portrays the mathematician, who was stripped naked by her murderers, in front of a Christian altar, symbolizing the conflict between her paganism and the then-emerging religion of Christianity.

her tragic death signaled the end of classical Greek mathematics—and the classical period in general—although some work in mathematics in Greece continued for several decades.

The sack of Rome in the fifth century certainly marked the beginning of the economic and cultural decline of Europe. Whatever mathematical and philosophical ideas may have arrived on the main part of the continent from Greece seem to have been lost.

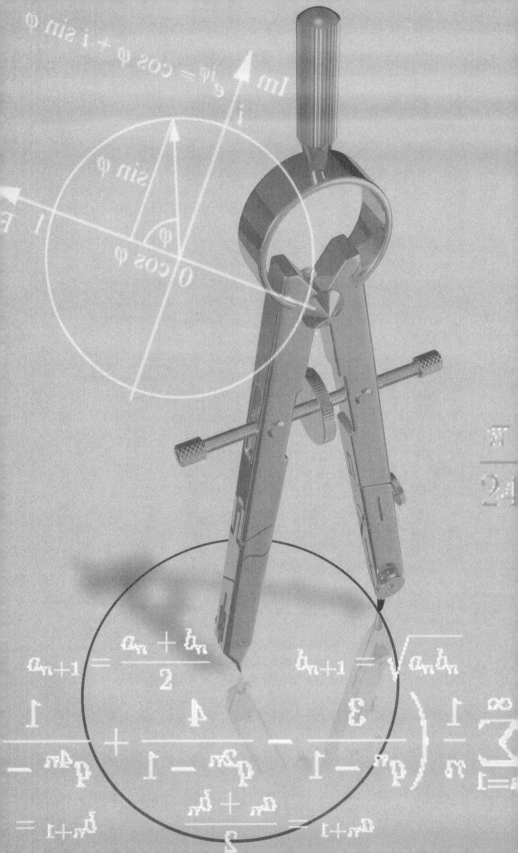

II

THE EAST

$$a_{n+1} = \frac{a_n + b_n}{2}$$

$$b_{n+1} = \sqrt{}$$

$$e^{i\varphi} = \cos\varphi + i\sin\varphi$$

$$\sum_{n=1}^{\infty} \frac{1}{n}\left(\frac{3}{q^n - 1} - \frac{4}{q^{2n} - 1} +\right.$$

$$a_{n+1} = \frac{a_n + b_n}{2}$$

$$a_0 = 1 \qquad b_0 = \frac{1}{\sqrt{2}} \qquad t_0 = \frac{1}{4}$$

FOUR

THE HOUSE OF
WISDOM

s Europe declined in the Middle Ages, an
Islamic empire was rising in the East. It included
Mesopotamia—the ancient site of some of the
earliest developments in mathematics—as well as the Arabian
Peninsula and regions stretching eastward beyond Persia and
westward to North Africa. As part of their cultural awakening,
the Arabs and Persians translated many classical works of the
ancient Greeks. For example, the Arab mathematician Thabit
ibn Qurra translated works of Apollonius, Euclid, and Archimedes. Incidentally, ibn Qurra's book *On the Sector-Figure*,
translated into Latin, would allow Newton to elaborate on

Caliph Haroun al-Rashid ruled the Arab Empire during its golden age in the late seventh and early eighth
centuries, and his court at Baghdad helped make the city a center of mathematical scholarship and
discovery. In this illustration, the emperor Charlemagne receives an ornate water clock that the caliph
had sent as a gift.

Apollonius's work almost two thousand years after the Greek mathematician's lifetime. In India mathematics also flourished during this period.

ARYABHATA

In 476, a date often associated with the demise of the Roman Empire, a mathematician named Aryabhata was born in India. His birthplace is believed to have been somewhere in central India. When he was twenty-three years old, Aryabhata wrote a book entitled the *Aryabhatiya*, which played a role in Indian mathematics somewhat akin to that of Euclid's *Elements* in the West. This treatise explained much of the arithmetic and calculations used in astronomy. It described powers of ten up to ten to the tenth—a huge number to comprehend at that time—and explained how to compute the square and cube roots of integers. Amazingly, it also provided the formula for the area of a triangle, rules for the sums of the terms in arithmetic progressions, explanations of geometric progressions arising from problems about compound interest, and work related to the solution of quadratic equations. These are surprisingly extensive results and not likely to have all been derived by one man—especially one so young. Most likely, the *Aryabhatiya* is a compilation of the work of many mathematicians, native and foreign, or a survey of the state of the art in Indian mathematics. Greek coins found in India indicate early trade relations, so, conceivably, the derivations in the *Aryabhatiya* may have originated elsewhere and been elaborated on by Aryabhata, his predecessors, and his contemporaries. Aryabhata also devised an estimate of π, which he described as follows:

> Add 4 to 100, multiply by 8, and add 62,000. The result is approximately the circumference of a circle of which the diameter is 20,000.[1]

Aryabhata's calculation implies an estimate of π equal to 3.1416—as good as the estimate provided by the Greek astronomer and mathematician Ptolemy. This number and the units used (ten thousands) further imply a connection between India and the Greek world of the early centuries CE—one that certainly existed through trade.

The Jantar Mantar observatory in New Delhi, India, constructed around 1724, contains thirteen astronomical instruments that employ technology developed centuries earlier, during the time of the mathematician Aryabhata (476–550).

Much of Indian mathematics was derived from astronomical interests. The Indians were always great astronomers, intertwining the study of the stars with religion as early as the second millennium BCE. For example, the four-thousand-year-old *Rig Veda* texts reveal sophisticated astronomical calculations, including a year composed of twelve thirty-day months. At such cities as Delhi and Jaipur, one can still visit the Jantar Mantar (calculation instrument) observatories developed under the reign of Maharaja Jai Singh II in the eighteenth century, but the large stone constructions with numerical markings designed to measure the movements of stars, planets, and the sun across the sky exhibit technology that existed at the time of Aryabhata. In fact, Aryabhata worked as an astronomer, mathematician, and teacher at early institutions of learning and research in India. Before his death in 550, he published a number of mathematical and astronomical works, providing definitions for the sine and cosine functions that form the basis of trigonometry, as well as calculations for Earth's rotation and revolution around the sun that differ from modern values by only .01 seconds and 3.33 minutes, respectively.

BRAHMAGUPTA

Half a century later India produced one of the greatest mathematicians who ever lived on the subcontinent. Little is known about the life of Brahmagupta, although we know that he was born in the year 598. A historian named Prthudakasvamin mentioned Brahmagupta in his ninth-century writings, referring to him as "the teacher from Bhillamala," which we now identify as Bhinmal, near Mount Abu in Rajasthan.[2] He is believed to have headed the Jantar Mantar in the city of Ujjain in central India through much of his life.

Brahmagupta wrote a number of books on mathematics, the most important of which was the *Brahmasphuta Siddhanta*. A *siddhanta* is a system in astronomy, written in Sanskrit as a book or manual. Such treatises were prevalent in India at this time, but Brahmagupta's was the most important because it was very complete, including many important mathematics and astronomy ideas having to do with calculation, arithmetic, and early trigonometry.

Brahmagupta did not possess good estimates of π—he used three and the square root of ten. He did, however, devise formulas for areas, such as that of an isosceles triangle and a quadrilateral, as well as rules for the solution of quadratic equations—including negative roots! The use of negative numbers as solutions to equations was pioneered by this brilliant

Brahmagupta (598–ca. 668), the pioneering Indian mathematician, discovered many solutions to problems of mathematics and astronomy. He is also credited with being the first to identify zero as a number.

mathematician. He also understood trigonometry and provided solutions that required the sine function.

As mentioned, Indian mathematicians were primarily interested in problems of astronomy. In particular, they wanted to know the average positions of heavenly bodies based on actual observations. Indian mathematicians working in an observatory measured the positions of the planets, the sun, and the moon over many months and years, recorded them, and then averaged all observations for each heavenly body. Based on his computations of the positions of heavenly bodies, Brahmagupta conjectured that the universe was 4,320,000,000 years ("revolutions of the sun") old.[3] There is no explanation of how or why he obtained this number from observations of the sky, but his estimate is strikingly close to the present estimate of 4.5 billion years for the age of our solar system.

Using the gnomon, an ancient astronomical device, Brahmagupta could also determine the latitude of his position from the shadow of the sun at noon during the equinox. However, because he was able to predict eclipses and do many astronomical computations that other mathematicians couldn't, he became conceited about his abilities and criticized his predecessors—especially Aryabhata—in his writing.

THE WORK OF BRAHMAGUPTA and other Indian mathematicians made its way to Arabia by 766, where it found fertile ground for growth. Arabia, the Arabic-speaking region extending roughly from central Asia to North Africa, flourished just as Europe was declining. By 775 the Indian texts had been translated into Arabic, and a few years later, Ptolemy's *Almagest* was also translated. Thus, by the dawn of the ninth century, the Arabs possessed knowledge of both the Greeks and the Indians who preceded them and were ready to make their own contributions to mathematics. During this "Islamic Golden Age," the center of the Arab empire was Baghdad, and its ruler was a caliph.

Haroun al-Rashid (ca. 763–809) was an enlightened ruler who brought art, culture, and ideas to his people and made Baghdad an enchanted, storied city—one thinks of the timeless anthology *Arabian Nights* and the heroic adventurer Sinbad the Sailor. Subsequently, Caliph al-Mamun, who ruled between 809 and 833, continued in the footsteps of al-Rashid and turned Baghdad into "the new Alexandria." Reportedly, one night the caliph had a dream in which Aristotle appeared and spoke to him. When he awoke, al-Mamun decided that he would bring to Baghdad all the works of the ancient Greeks. These texts were philosophical, astronomical, and mathematical—Euclid's *Elements* being foremost among them. The caliph also founded the House of Wisdom, which he modeled after Plato's Academy. Important thinkers of this prestigious body included Omar Khayyam, a mathematician and famous poet, and a man named Mohammad ibn Musa al-Khwarizmi, from whose name we get the modern word *algorithm*.

AL-KHWARIZMI

Al-Khwarizmi (d. ca. 850) came to Baghdad from a region in present-day Azerbaijan. Unfortunately, nothing is known about his life. Presumably, he had access to the work of Brahmagupta, and his own voluminous oeuvre seems to have been based on Indian mathematics. We don't know exactly which works he brought to Baghdad, but all indications point to Brahmagupta's great treatise, *Brahmasphuta Siddhanta*. In particular, al-Khwarizmi's book *Concerning the Hindu Art of Reckoning* is believed to be based on the work of Brahmagupta. Beginning with the great Indian mathematician's work on quadratic equations, al-Khwarizmi launched a project on the algebraic solutions of equations so extensive that today we consider him a founder of algebra. In addition to writing the first comprehensive book on algebra, he introduced Hindu numerals into Arabic mathematical literature. These Arabic-Hindu numerals would later find their way into Europe, along with his algebraic methods and astronomical calculations.

Al-Khwarizmi expanded the algebraic work of Diophantus and popularized algebra

Muhammad ibn Musa al-Khwarizmi popularized the field of algebra with his book *Al Gabr Wa'l Muqabala*, published in 830. His likeness appears on this Russian commemorative stamp, issued approximately twelve hundred years after his birth.

The House of Wisdom, founded by Caliph al-Mamun, was an active library and intellectual center in Baghdad until its destruction by Mongol invaders in 1258. This illustration of a group of scholars in an Islamic library comes from the Maqamat of al-Hariri, an illuminated manuscript of the thirteenth century.

by showing, in words rather than symbols, how equations could be set up and solved in a complete and systematic way. By simplifying the highly theoretical methodology of the Greeks, Indians, and Babylonians, he thus set a firm and practical foundation for the field of algebra for generations to come. Al-Khwarizmi's major work on algebra, *Al Gabr Wa'l Muqabala* (from which the modern word *algebra* is derived), is a guide to solving equations, especially quadratic equations, in practical, real-world scenarios. The word *al-gabr* means "restoration" or "completion" and implies moving elements from one side of an equation to the other. *Muqabala* means "reduction" or "balancing"—canceling terms when they appear on both sides of an equation. The book's first six chapters show how to systematically solve different kinds of equations, whereas the rest of the book uses geometry to demonstrate the validity of the methods presented in the first six chapters. The pragmatism of the first part resembles the practical Babylonian approach that we see in clay tablets from ancient Mesopotamia. Later Arabic texts even referred students to methods explained in *Al Gabr Wa'l Muqabala,* suggesting that the book had become a classic in the heyday of the Arabic Empire. After being translated into Latin, *Al Gabr* made its way to Europe.

THE HOUSE OF WISDOM was home to a number of impor-
tant mathematicians, astronomers, and scholars. Arabia is conveniently
located between the world of ancient Greece and that of India and East
Asia. As a result, the Arab empire borrowed from both West and East,
building upon the straight-line geometry developed by the Greeks while
employing trigonometric sine tables created by the Indians. For example,
the Arab astronomer al-Battani (ca. 850–929), known in the West as
Albategnius, used the Hindu sine function in his book *On the Motion
of the Stars,* which enabled the medieval European astronomers who followed him to use trigonometry in calculating motions of heavenly bodies.
He is credited with deriving the relation that the tangent of an angle is
equal to its sine divided by its cosine. Computations were accurate to
eight decimal places, and angles were accurate to a quarter of a degree.[4]

By the tenth century, Arab mathematicians derived further trigonometric identities, such as the law for the doubling of an angle, $\sin(2x) =
2\sin(x)\cos(x)$. But the heyday of Arab mathematics came in the eleventh
century. An Arab mathematician named al-Karkhi, who is known to
have lived around 1029, extended the work in Diophantus's *Arithmetica*
to equations of higher orders and ones in which coefficients and solutions
were not restricted to being rational numbers. In addition to al-Karkhi,
that period saw the emergence of Ibn Sina (980–1037), known in the
West as Avicenna. One of the most prominent scholars of this period, he
made contributions not only to mathematics but also to medicine and
philosophy.

Avicenna's contemporary al-Biruni (973–1048) traveled widely. His
famous book, *Indica,* described Indian science and showed how a
nonagon—a polygon with nine sides—could be inscribed in a circle,
using the trigonometric formula for the cosine of 30 degrees to show
that the problem of inscribing the nonagon in a circle was equivalent to
solving the equation $x^3 = 1 + 3x$. (He solved the problem to an amazing
accuracy of six decimal places.) *Indica* also included a discussion of the
Indian heliocentric theories of Aryabhata and Brahmagupta, noting that
the idea of Earth revolving around the sun and rotating about its own
axis were consistent with his astronomical calculations and, thus, could
not be refuted.

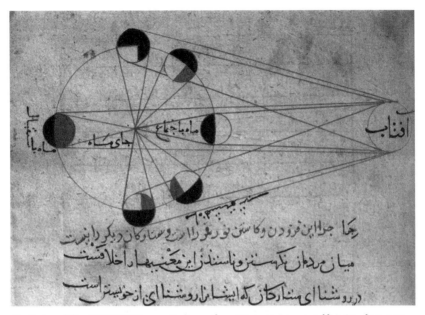

Abu Rayhan al-Biruni (973–1048) was a Persian mathematician, astronomer, and linguist whose many important books include the astrological treatise *Kitab al-Tafhim*, which includes this illustration of the phases of the moon.

An Egyptian mathematician named Ibn Yunus (ca. 950–1008) extended trigonometry even further by introducing a powerful formula, $2\cos(x)\cos(y) = \cos(x + y) + \cos(x - y)$, which translates a product into a sum. Because this formula turns the product (on the left side of the equation) into a sum (on the right), it *lowers the order of computation* (because adding is easier than multiplying, and subtraction is easier than division). Ibn Yunus's formula thus offered a tool for simplifying calculations. This prelogarithmic method for aiding computation was given the name *prosthaphaeresis*, which in Greek means "addition and subtraction." Later, and until the advent of the modern calculator, logarithms were used to carry out multiplications as sums and divisions as subtractions through a slide rule and similar devices.

OMAR KHAYYAM

Omar Khayyam (1048–1123) was a Persian mathematician and poet, and it was he who took al-Khwarizmi's pioneering ideas on algebraic equations and solutions to a new level. In his seminal work, *Algebra*, this poet-mathematician showed how to solve quadratic equations both algebraically and geometrically. He studied cubic equations, revealing the steps to their geometric solution, but he assumed (falsely) that a purely algebraic solution to cubic equations was not possible. Geometrically, he solved cubic equations using a method known to the ancient Greeks—the intersections of conics—but he generalized this method to any given arbitrary cubic equation with positive roots. Because we live in three-dimensional space, Khayyam was stumped in his attempts to solve equations of a higher order than three, not being able to envision the space in which such equations lived. He was equally hampered in his work by the fact that he and his contemporaries did not understand that negative solutions to equations have meaning. Thus he worked with only positive coefficients. The recognition that negative numbers are meaningful would come later, as mathematics matured, as would the so-called imaginary numbers.

In general, the Arabs followed the Indian approach, which emphasized algebra and trigonometry over geometry, the realm of the ancient Greeks. Khayyam's great genius was that he was able to use and understand both approaches. His thinking along the parallel lines of geometry and algebra foretold Descartes's unification of these two mathematical fields in the seventeenth century. Presciently, Omar Khayyam wrote, "No attention should be paid to the fact that algebra and geometry are different in appearance. Algebras are geometric facts which are proved."[5] The connection between geometry and algebra, however, would not be pursued to a great extent by the Arabs. Regarding algebraic solutions of higher-order equations, Khayyam conceded, "they are impossible for us and even for those who are experts in this science. Perhaps one of those who will come after us will find them."[6]

ATTACKING THE FIFTH POSTULATE

Omar Khayyam and other Arab mathematicians had a near obsession with Euclid's fifth postulate, which states that if two lines that, intersecting a third line, make two interior angles whose sum is less than the sum of two right angles—that is, 180 degrees—then the lines, if extended far enough, will intersect on the side of the third line on which the sum of the two angles is smaller than the sum of two right angles. To these mathematicians, the fifth postulate was not an axiom, but rather something that, like a theorem, was provable based on the previous four postulates Euclid had put forward. But no one could prove the fifth postulate.

The Arab mathematician Alhazen (ca. 965–1039), who had written the highly influential *Book of Optics*, attacked what is now known as the "parallel postulate." After constructing a tri-rectangular quadrilateral, he thought he could prove that the fourth angle must be a right angle. What he did not realize was that he was operating on an assumption that, itself, was equivalent to Euclid's fifth postulate—the result he was trying to prove.

Omar Khayyam criticized Alhazen's attempted proof, saying that Aristotle had ruled out the viability of motion in geometry. In his own attempt at a solution to this problem, he constructed a quadrilateral whose two sides were perpendicular to the base, and concluded that the two interior angles formed by the fourth side must be equal to each other. Referencing the Aristotelian principle that two converging lines must intersect, he ruled out the possibility that the two angles could be either acute or obtuse and concluded that they were right angles. However, the Aristotelian principle he invoked is, in fact, equivalent to Euclid's fifth postulate, so both mathematicians failed in their respective "proofs."

A century later an Arab mathematician named Nasir al-Din al-Tusi (1201–74) mounted the most consequential attack on the problem of the fifth postulate. Tusi was the astronomer of the Mongol ruler Hulagu Khan, grandson of Genghis Khan and brother of Kublai Khan. Tusi proceeded from the same geometrical starting point as his predecessors, the quadrilateral, and his "proof" also depended on

Euclid's assumption, but his analysis would be useful in later centuries in the development of non-Euclidean geometries.

AL-KASHI

Major Arab and Muslim contributions to mathematics lasted into the fifteenth century. The last important Muslim mathematician of historical times was Jamshid al-Kashi (ca. 1380–1429). Al-Kashi was born in Kashan, central Persia. Around the time of his birth, the Mongol empire was expanding and conquering parts of Persia, and life was difficult in the realm. Al-Kashi, who early in life showed great promise in mathematics, became a wandering scholar eking out a living by teaching mathematics in the villages and towns in the region. We know that on June 2, 1406, he observed and recorded an eclipse of the moon, and his note has been used to date other events with accuracy. He became a respected astronomer after making many observations of the sky and publishing a book in 1407, the abbreviated title of which was *The Stairway of Heaven*. It dealt with measuring angular distances between heavenly bodies and performing calculations in astronomy.

His renown brought al-Kashi to the attention of Ulugh Beg, a grandson of the Mongol emperor Tamerlane. Ulugh Beg was the ruler of the ancient city of Samarkand, in present-day Uzbekistan, and he was

Ulugh Beg (1394–1449) is pictured alongside the observatory he built in Samarkand, the modern-day capital of the Samarqand Province of Uzbekistan. This Russian commemorative stamp also features the date of his initial calculation of the sidereal year: 1437.

very interested in science and mathematics. He also had great plans for the city, including founding a university and building an observatory, so he invited al-Kashi to join him there, along with other mathematicians and thinkers. Al-Kashi accepted the invitation and became the most prominent mathematician and astronomer in Samarkand, the "new Baghdad."

Al-Kashi specialized in the solution of quadratic equations. He seemed to relish doing long calculations, and he is credited with bringing decimal notations and computations, ideas that were later further pursued in Europe, to the fore. He computed roots of equations to an accuracy never before reached by any of his predecessors. He not only established the use of decimal fractions but also computed π to fourteen decimal places: 3.14159265358979.

Two centuries before the French mathematician and physicist Blaise Pascal published his treatise on the triangle of binomial coefficients, now called Pascal's triangle, al-Kashi discussed its properties in his *Key to Arithmetic*. In fact, the triangle had been discovered in China a century earlier. Al-Kashi died in Samarkand in 1429 and was sorely missed by Ulugh Beg, who considered him the greatest scientist of the time. His death signaled the final decline of Arab and Islamic mathematics and the ascendance of the West.

FIVE

MEDIEVAL CHINA

 elieved to be as old as the Babylonian and Egyptian cultures, the Chinese civilization is far more ancient than that of Greece. Although dates are uncertain, we know that mathematics has been studied in China for thousands of years. An early Chinese mathematical text of an uncertain date contains work on fractions and descriptions of the properties of right triangles, as seen in the mathematics of ancient Greece. It also recounts a debate between a ruler and one of his ministers in which the minister tells the ruler that the art of numbers originates in two figures: the circle and the square. The square represents the earth, and the circle represents heaven. This example of mathematical mysticism is reminiscent of the Pythagorean order's conviction that "God is number."

The Beijing Ancient Observatory, completed in 1442, contains many pretelescopic astronomical instruments, such as this bronze armillary sphere, used to calculate the coordinates of celestial bodies.

THE MAGIC SQUARE

Part of the magic of numbers manifested itself in the very ancient Chinese practice of constructing magic squares, which is believed to have originated before the first century CE—some say much earlier. The Chinese seem to have been the first to invent the magic square and to construct many such squares for their own entertainment and for mystical uses—much as they used compass needles in the ancient discipline of feng shui, the art of orienting buildings and furniture for maximum spiritual harmony. Magic squares were believed to have mystical qualities, and in ancient times some people wore them around their necks or kept them displayed on walls or doorways. Below is one of the first magic squares discovered in China:

4	9	2
3	5	7
8	1	6

THE MYSTIC TABLET.[13]

This magic square, or Lo Shu square, is of Tibetan origin. In the center of the tablet are the numerals of the magic square, written in Tibetan characters and emblazoned on the back of a turtle.

Note that the sum of all rows, columns, and diagonals is 15. This particular square also has mythological origins. According to Chinese tradition, a turtle from the River Lo brought this magic square to humankind in the days of Emperor Yii, about whom there were many legends. This magic square was given the name Lo Shu, which means "river map," although its derivation is obscure.

In early antiquity—certainly before the first century—the Chinese began to use rods of bamboo, iron, or ivory to perform calculations. Officials often carried such rods with them to calculate taxes and other financial data. These rods were manipulated with immense dexterity and were described by onlookers as "flying so quickly that the eye couldn't follow their movement."[1] The abacus seems to have been derived from this ancient Chinese system of calculation, though it is now believed that, because the rods were so efficient, they were used for much longer than originally thought, and the abacus emerged only relatively late in Chinese history.

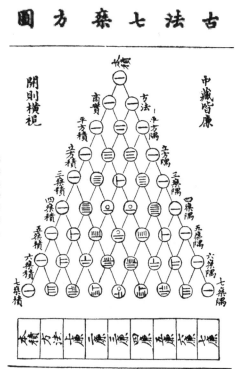

Chinese mathematicians in the Middle Ages used counting rods to perform various calculations. This illustration of Yang Hui's triangle (a triangular arrangement of the binomial coefficients, later called Pascal's triangle) uses counting-rod numerals within its structure.

NINE CHAPTERS

The most important Chinese mathematical text—and one of the oldest in existence—is the "Nine Chapters on Mathematical Art," a treatise dating from about 250 BCE, although some believe it is much older and may have been initiated by a series of mathematicians as early as the tenth century BCE. In contrast with the abstract mathematics of the Greeks, the Nine Chapters centers on practical issues of a mathematical nature, such as problems of taxation, agriculture, and engineering. The book provides rules for the areas of triangles, circles, and trapezoids. There is also a discussion of how many equations are needed to solve a problem with a given number of unknowns. For example, one problem presents four equations with five unknowns (which we now know is impossible, since the number of equations must equal the number of unknowns if we want unique solutions).

A commentator on the Nine Chapters who lived in the third century CE estimated π to be 3.14 by using a regular polygon of 96 sides, and later obtained an even better approximation, 3.14159, by using a polygon of 3,072 sides. The polygons allow us to calculate areas, and when these areas are added, an estimate of the area of the circum-

scribing circle is obtained, leading to an estimate of π. Estimating π seems to have been a major occupation of Chinese mathematicians, who came up with better and better estimates as time passed. The Nine Chapters also gave us the famous "Chinese remainder theorem," which is used in number theory, relating equations and their solutions with properties of numbers.

The opening page of chapter 1 of "Nine Chapters on Mathematical Art," one of the oldest mathematical treatises in existence.

LI ZHI

Li Zhi (sometimes spelled Li Chih) was an important mathematician of the thirteenth century. He later changed his name to Li Ye to avoid confusion with the third emperor of the Tang dynasty. His father was an official of the Jurchen empire, which encompassed Manchuria and northern China. Li Zhi was born in Beijing, the Jurchen capital, in 1192. When he was a teenager, the Mongols under Genghis Khan attacked northern China and took Beijing, so his family sent him to attend school in Hopeh Province.

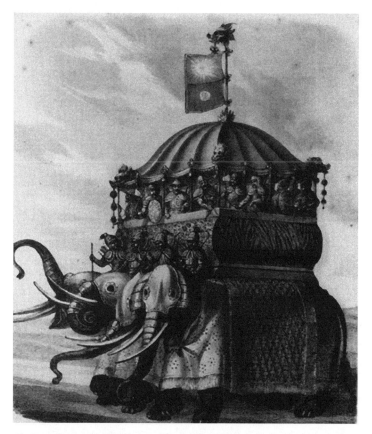

The Mongol emperor Kublai Khan—who, allegedly, would ride into battle in an armored fortification atop four elephants, as depicted in this 1874 illustration—was a great champion of mathematics who twice offered Li Zhi a position in his government.

When he finished his education, Li Zhi took the examinations for civil service in 1230. He was about to take a position as registrar of the district of Kaoling, but persistent Mongol attacks prevented him from assuming the promised job. In a strange twist of fate, he was then offered a better position in the civil service: that of governor of the Jun prefecture in Henan Province. The job didn't last long—after continued Mongol attacks, he barely escaped certain death when a colleague managed to convince a Mongol soldier to let him live.

For more than a decade, Li Zhi lived in abject poverty in yet another part of China to which he had to escape: Shanxi Province. Still, he was so gifted that even under such dire circumstances, he managed to complete a major work in mathematics, the *Ce Yuan Hai Jing* (Sea Mirror of Circle Measurements), in 1248. Teaching mathematics and distributing his book eventually improved his economic situation, so he moved back to Hopeh Province, where he had hoped to live in peace. In 1257 Kublai, grandson of legendary Mongol conqueror Genghis Khan, sent emissaries to Li Zhi—by then a famous mathematician—asking him for advice on restructuring the civil-service examinations. Apparently, he liked Li Zhi's proposal, because in 1260, after he became the Mongol leader Kublai Khan, he offered Li Zhi a top administrative job controlling his vast empire.

Already in his sixty-ninth year, Li politely declined the offer, saying he was too old to be a top administrator. Four years later Kublai Kahn appointed Li Zhi to his newly established scientific academy, but the aging mathematician apparently felt unwell and resigned shortly after joining the Mongol academy. He returned to his old home in a peaceful mountain region of China, where he lived out his last years and died in 1279, at the age of eighty-seven.

Li Zhi's book included 170 problems that had never before been solved. He was interested in the relationships between the sides of triangles and the radii of the inscribing or inscribed circles. Such analysis, usually involving the Pythagorean theorem—which is now believed to have been discovered independently in China in early antiquity—led to equations of varying degrees. Li Zhi therefore studied equations as high as the sixth order. Such an equation is obtained when one relationship about triangles involving an unknown quantity is inserted into another. Repetitive application of such a procedure quickly raises the order of the unknown in the equation. While Li Zhi showed how to set up equations in order to solve a problem, he did not describe his methods of solution. This focus on equations of a higher and higher order would become very important as Europe emerged from the Dark Ages and built upon the mathematics of antiquity and the Near East.

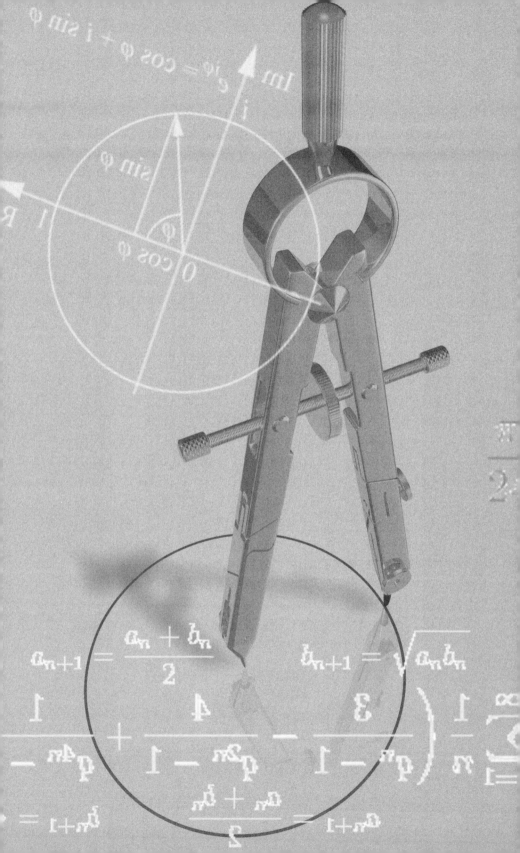

III

RENAISSANCE MATHEMATICS

SIX

ITALIAN SHENANIGANS

A ccording to historian of mathematics Carl Boyer, the main problem with the transmission of mathematical ideas in historical times was language. The Arabs, according to Boyer, made their big breakthrough when they were finally able to translate into Arabic the important foundational work of the Greeks. The Europeans would not be able to penetrate the Arab language barrier for another three hundred years. When this finally happened, it changed the world.

FIBONACCI

As interest in astronomy grew in the twelfth and thirteenth centuries, European scholars began to study Arabic so that they could understand the mathematical works of the Arabs. It was thus that al-Khwarizmi's work on algebra was brought to the attention of European mathematicians.

Leonardo da Vinci's ca. 1487 drawing *Vitruvian Man* has come to symbolize the creative, inventive spirit of the Renaissance. Da Vinci's affinity for mathematics informed his friendship with Luca Pacioli, another great Italian mathematician of the period.

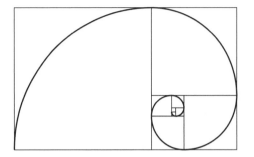

A Fibonacci spiral is a series of connected quarter-circles drawn inside an array of squares whose dimensions correspond to the numbers in the Fibonacci sequence.

Prime among these new European mathematicians was Leonardo of Pisa (ca. 1170–1250), better known as Fibonacci (son of Bonacci). Fibonacci was the first European mathematician to use the Hindu-Arabic numerals we use today. He found these numerals in the works of al-Khwarizmi and other Arab scientists, and thanks to Fibonacci's writings, these numerals—or, rather, the version of them that evolved in Europe—replaced the very cumbersome and inefficient use of letters as numbers. Using the new numerals allowed Fibonacci to formulate his famous sequence in simpler notation (i.e., 1, 1, 2, 3, 5, 8, 13, 21 . . . rather than I, I, II, III, V, VIII, XIII, XXI . . .). In Fibonacci's sequence—first derived as a series of numbers representing an idealized scenario in which a pair of rabbits mates and produces a new pair of rabbits every month—each number represents the sum of the two numbers immediately preceding it. As it turns out, the ratio between each term and its predecessor approaches 1.618 . . . , the golden ratio, which appears commonly in nature and in art.

How did he do it? Relative to most mathematicians, Fibonacci had an advantage: his father, Guglielmo Fibonacci, was a successful merchant and trader who traveled to North Africa and often took his son with him. To aid his father, Leonardo learned Arabic and thus had a much better handle on the language than other European mathematicians of the time. He also gained access to many Arabic texts and assimilated their ideas into his work. The rediscovery and blending of ideas through the Arabic texts and translations of ancient Greek mathematics allowed Fibonacci to single-handedly jump-start European mathematics.

As the son of a wealthy family, Fibonacci was able to travel easily—and his father's trade relations in the Mediterranean helped him there as well. As a young adult, Fibonacci embarked on a multiyear adventure, visiting many ports in the Mediterranean and studying with Arab mathematicians.

He also became proficient in the use of the Hindu-Arabic numerals. In the year 1200, when he was around thirty years old, he returned to his native Italy. Two years later he published his masterpiece, the *Liber Abaci* (Book of Calculation), which promoted the use of the Hindu-Arabic numerals, discussed his explorations in number theory and geometry, and introduced his famous sequence. The book described the "nine Indian figures"—the digits 1 through 9—and the symbol that designates the concept of *zephirum* (zero): 0. Fibonacci did not use decimal fractions, however; they would appear later in European mathematics.

The *Liber Abaci* also described the extraction of roots in the Arab tradition of al-Khwarizmi, and covered problems in finance and trade, including foreign-exchange calculations. An example of the kind of problem that appears in the *Liber Abaci* is the following:

> Seven old women went to Rome; each woman had seven mules; each mule carried seven sacks; each sack had seven loaves; and with each loaf were seven knives; each knife had seven sheaves.

The question was to find the sum of all of these: women, mules, sacks, loaves, knives, and sheaves. A very similar problem appears in the Egyptian Ahmes papyrus, dating from about 1600 BCE. The next example, from the *Liber Abaci*, gave rise to Fibonacci's sequence:

> How many pairs of rabbits will be produced in a year, beginning with a single pair, if in every month each pair bears a new pair, which becomes productive from the second month on?[1]

Fibonacci had an interesting relationship with the Holy Roman emperor. Frederick II had been crowned

A page from Fibonacci's monumental work *Liber Abaci* (1202), which advocated the use of Hindu-Arabic numerals.

king of Germany in 1212. Eight years later, the king was elected Holy Roman Emperor. In November that year, the pope crowned him in a majestic ceremony at the Cathedral of St. Peter in the Vatican.

As Holy Roman Emperor, Frederick sided with Pisa in its rivalry with Genoa and was thereafter very popular with the Pisans. He became familiar with the mathematical work of Pisa's native son because Fibonacci had an extensive correspondence with the scholars and mathematicians already in Frederick's court. The emperor's imperial court met in Pisa in 1225, and some of the city's dignitaries, including the Fibonaccis, were invited to a banquet. Court mathematicians issued problems as a contest, and Fibonacci correctly answered a number of them, gaining the increasing respect of the emperor. Soon afterward Frederick invited the brilliant Fibonacci to join his court. Fibonacci accepted and eventually became the emperor's favorite mathematician. In addition to being paid by the emperor, the mathematician's native city established a salaried position that paid him a stipend for consulting on issues of administration and taxation, and on any problem requiring mathematical analysis. Furthermore, it was through Fibonacci's influence that Frederick II founded the University of Naples to promote his vision of a good Italian education. This university is still referred to by Italians as Federico II, in recognition of the thirteenth-century emperor's generosity.

Fibonacci wrote several books on mathematics. In *Flos* (*The Flower*), published in 1225, he provided solution methods for sophisticated equations that resembled those of Diophantus of Alexandria and, hence, went far beyond the practical-minded work of al-Khwarizmi and other Arab and Hindu mathematicians. The book also provided groundbreaking methods for the solution of cubic equations. For example, he gave a very good approximation to a solution of the equation $x^3 + 2x^2 + 10x = 20$, a problem that had been issued as a challenge by a mathematician named Johannes of Palermo but that originated in the work of Omar Khayyam. One of his books, now lost, was a commentary on Euclid's *Elements* and included a treatment of irrational numbers, which the Europeans had not yet addressed following their discovery by the ancient Greeks. Fibonacci discussed them from a purely computational viewpoint without addressing the philosophical difficulties involved.

Fibonacci's seminal work, *Liber Abaci,* was reissued many times and became a major European text in mathematics well into the nineteenth century. Through his famous sequence and his introduction of the efficient Hindu-Arabic numerals to European mathematics, Fibonacci achieved immortality as the leading mathematician of his time.

THE INVENTION OF PERSPECTIVE

As Euclid's *Elements*—one of the first books ever to be printed on the new printing presses in Venice after their invention in 1440—and al-Khwarizmi's *Algebra* were finding wide distribution and success in Europe during the fifteenth and sixteenth centuries, their ideas were further pursued in the same region of northern Italy that was flourishing culturally and economically under the influence of the Venetian mercantile empire. It has also been asserted that the fall of Constantinople to the Turks in 1453 led to mass migrations of educated Byzantine citizens to Italy and that these refugees brought with them many intellectual works of the East, including Arab mathematical writings. Thus, with the Byzantine decline and the mechanization of bookmaking, Italian and other European mathematicians came to possess the priceless works of their predecessors and were in a position to expand and extend these earlier achievements.

In the sixteenth century the pursuit of geometry in Italy and Germany was buoyed by the invention of perspective in Renaissance painting. Medieval artists had a general idea of perspective, representing distant objects as smaller than objects in the foreground, but there was no governing system of measurement that determined the positions and sizes of various elements. In the fourteenth century, however, the geometrical portrayal of vision in Alhazen's *Book of Optics* caught the attention of Renaissance painters such as Filippo Brunelleschi (1377–1446) and Leon Battista Alberti (1404–72).

In Germany, Johannes Werner (1468–1522) and artist Albrecht Dürer (1471–1528) pioneered the technique. Werner worked on conic sections and made futile attempts to solve the persistent doubling-of-the-cube problem of classical Greece. His work on conic sections, however, led to results that foreshadowed the idea of perspective.

Albrecht Dürer's symbolic 1514 engraving *Melencolia I* provides ample evidence of the artist's interest in mathematics. Note the tools of geometry and architecture at the feet of the seated woman, the polyhedron with an unknown number of sides, and the magic square at the top right.

Intrigued by the concept of fixing elements to a mathematical grid in order to convey the illusion of depth, Dürer identified a fixed point on a circle and then let the circle roll along the circumference of another circle, which generated an epicycloid. He was not a mathematician, so he lacked the tools to produce a precise analysis, but he was able to make projections of helical curves onto a plane, producing spirals, which he incorporated into his art. He also produced an approximation of a nonagon—a polygon with nine sides.

In many ways Dürer was ahead of his time, as mathematics was not yet advanced enough to allow him to create art with sophisticated mathematical tools. Instead, like many mapmakers of that era, Dürer used projections that were not understood by mathematicians.

LUCA PACIOLI

Franciscan friar Luca Pacioli (1445–1517) explored algebra and geometry, including the idea of linear proportion, which was increasingly being incorporated into works of art. Pacioli was born in the town of Sansepolcro, which lies in the Apennine Mountains of central Italy, about halfway between Perugia and Florence. For some reason he was not raised by his parents; rather, he was raised by the Befolci family, who lived in Sansepolcro. Not far from their home was the studio of the famous artist Piero della Francesca. This painter pioneered the use of perspective in art, and Pacioli is believed to have spent time in the artist's studio, learning from the master about perspective as a child. When he became a young man, Pacioli tutored the three sons of a wealthy Venetian family named Rompiasi, who were living on the island of Giudecca—one of Venice's best neighborhoods. In Venice the young tutor continued his mathematical education on his own, reading mathematics books and deriving his own results. He wrote a book on arithmetic and dedicated it to the three young boys he was tutoring. When he finished his job educating the boys, Pacioli left Venice for Rome, where he made strong contacts in Catholic circles and joined the Franciscan order, becoming a friar.

This 1495 portrait of Luca Pacioli, attributed to Jacopo de'Barbari, depicts the mathematician demonstrating one of Euclid's theorems. The figure in back is Pacioli's student Guidobaldo da Montefeltro. A rhombicuboctahedron (which has eight triangular and eighteen square faces), half filled with water, hangs from the ceiling.

Pacioli loved to travel. In 1477 he left Rome and traveled to Perugia, where he taught mathematics at the local university until 1480. Then he moved again, teaching at the University of Naples and later the University of Rome. Through his ecclesiastical connections, he met Federico da Montefeltro, whom the pope had recently made the duke of Urbino. An enlightened ruler interested in education and science, Montefeltro invited the wandering friar to tutor his son Guidobaldo, who would become the last ruler of Urbino from the Montefeltro family. When this position ended, Pacioli again moved to Rome, before eventually returning to Sansepolcro. He had achieved fame as a mathematician by that time, inciting jealousies from lesser but more powerful individuals in his hometown.

In 1496 another invitation arrived, from an unexpected source. Ludovico Sforza, the enlightened duke of Milan, was building up his city's cultural institutions to rival those of all other European cities. Sforza had brought to Milan the great painter, sculptor, inventor, engineer, architect, Renaissance man, and genius Leonardo da Vinci (1452–1519) after Leonardo had written him his now-famous letter in which he offered to effect several great engineering projects in Milan and mentioned, as an aside, that he could also paint. In Milan, Leonardo painted some of his best works of art, including the *Virgin of the Rocks* and *The Last Supper*. One of his many important projects in Milan was the design of the dome for the Milan Cathedral. It has been surmised that Leonardo, with his interest in mathematics, had suggested to Ludovico that he invite Luca Pacioli to his court. When Pacioli arrived in Milan, he and Leonardo became very close friends. They spent many hours together discussing the two topics that consumed them both: art and mathematics.

Two years earlier Pacioli had completed a major book on mathematics, *Summa de Arithmetica, Geometrica, Proportioni et Proportionalita*, a collection of his own and his predecessors' arithmetical, geometric, and algebraic work. It contained few attributions but was clearly influenced by al-Khwarizmi's *Algebra,* which had been in print for three decades by that time. Since he had received instruction in the terminology of commerce, his book also included work on double-entry bookkeeping, which was an Italian invention of the time that revolutionized business practices and inaugurated the modern field of accounting.

Like Omar Khayyam, Pacioli believed that cubic equations could only be solved geometrically and that an algebraic solution was something that mathematicians of the future might achieve. But that future was near, and it belonged to Pacioli's own countrymen. In Milan he spent time on geometry and produced a book, *De Divina Proportione*, that expounded on the golden ratio. The book revealed the relationship between polygons and three-dimensional solids and the ways in which various proportions follow the golden ratio, often called phi. The beautiful figures within the book were drawn by none other than Pacioli's friend Leonardo da Vinci. Leonardo himself used mathematical concepts in his writing. In his notebooks we find constructions of regular polygons and ideas about centers of gravity. In art he was a pioneer of the mathematical use of perspective.

In 1498 the French monarch Louis XII declared that Milan belonged to France, but the reigning duke refused to abdicate, so the French army attacked the city the following year. Ludovico had to flee his duchy, and in December 1499 Luca Pacioli and Leonardo da Vinci escaped together after French troops began occupying Milan. They stopped at Mantua, where they were the guests of Marchioness Isabella d'Este, and from there they continued to Venice. After a period of time in Venice, Pacioli and da Vinci continued to Florence, where they shared a house. Both men remained in Florence for several years, spending time away from the city teaching at various universities. Around this time, Pacioli began to work with another Italian mathematician, Scipione del Ferro, who plays a major role in the next significant development in Italian mathematics.

After further travels to Perugia and Rome, Pacioli returned to his native Sansepolcro, where he died in 1517, leaving behind an unpublished book about recreational mathematics that made frequent references to his famous friend, Leonardo da Vinci.

TARTAGLIA

With the transition from medieval to Renaissance Europe, the mathematics of the ancient Greeks, Indians, Arabs, Egyptians, and Mesopotamians converged in Europe—Italy, in particular—through the work of Fibonacci and his contemporaries. Finally, European mathematics came of age. With the solution of equations, algebra was mastered by European mathematicians, and new advances were on the horizon. Four Italian mathematicians played key roles in the development that followed.

Niccolò Fontana, known as Tartaglia (ca. 1500–59), was born in Brescia to an unmarried mother who lived in poverty. In 1512, when he was twelve years old, Niccolò's fortunes went from bad to worse. French troops led by Gaston de Foy attacked Brescia, and the boy was severely deformed after a French soldier in the invading army slashed his face with his sword. Niccolò's mother nursed him back to health, but his lips were so badly cut that he stuttered throughout the rest of his life and was thereafter known as Tartaglia (the stammerer).

As a young adult, Tartaglia moved to Venice and pursued the life of an aspiring mathematician. Tartaglia translated the treatises of Aristotle, Euclid, and other Greek mathematicians into Latin. In the meantime, a professor of mathematics at the University of Bologna named Scipione del Ferro (1465–1526), with whom Pacioli had worked, made a stunning discovery: he found a way to solve cubic equations. Nasir Adin al-Tusi had previously made some progress in understanding some of these equations, but no general formula had been known—the Arabs who had studied the

Niccolò Fontana, known as Tartaglia (the stammerer), wrote his solution for solving cubic equations as a poem.

cubic-equation problem could not come up with a good general way of solving it.

In 1526, while Tartaglia was working on this same problem in Venice, del Ferro died in Bologna. Just before his death, however, he revealed his big secret to a mediocre student named Antonio Maria Fior. What del Ferro had discovered was a way to solve cubic equations that contain no x^2 terms and in which all the coefficients are positive numbers.

Fior understood that he possessed a very powerful formula that could allow him to make large sums of money in competitions, so he moved to Venice and challenged other mathematicians to solve cubic equations. In 1535 Fior challenged Tartaglia. Each contestant was to provide his opponent with thirty problems to solve. All the equations Fior gave Tartaglia were in a form that he thought he knew how to solve using del Ferro's method—equations of the form $x^3 + px = q$—but he was ultimately unprepared for the wide variety of cubic equations that Tartaglia gave him.

Tartaglia, who managed to solve some of the cubic equations by educated trial and error, beat Fior because the latter had misunderstood del Ferro's formula and obtained wrong answers. Having pondered the problem of finding a general formula for the solution of cubic equations, Tartaglia gleaned from Fior's errors something about the mysterious formula.

During the night of February 12, 1535, Tartaglia managed to derive his own general formula for solving cubic equations—he had figured out del Ferro's secret. Tartaglia wrote his solution method as a poem, and instead of the x we use for an unknown quantity in mathematics today, he used *cosa*, the Italian word for "thing" (plural: *cose*). It began as follows:

> *Quando che'l cubo con le cose appresso*
> *Se agguaglia a qualche numero discreto . . .*
> [When the cube with the "things" is equal to a number . . .]

Because the word *cosa* represented the unknown in an equation, eventually all mathematicians concerned with solving equations in sixteenth-century Italy became known as cossists. This terminology originated with Luca Pacioli, who abbreviated *cosa* as *co*; *censo*—the square of the unknown *cosa*—as *ce*; and *aequalis*—the equal sign—as *ae*. Using Pacioli's notation, for example, the equation $5x^2 = 7x$ might be written as $5ce\ ae\ 7co$. But Tartaglia, who favored verse, did not abbreviate *cosa* and other elements of his analysis.[2]

Having derived a powerful method of solving cubic equations, Tartaglia was able to make a good living by taking part in public competitions of equation solving, which were held in piazzas and other locations in Venice. His discovery also brought him many pupils who wished to learn his methods, and he was offered a number of university positions. Soon word of his success reached a man named Cardano, who lived in Milan.

GIROLAMO CARDANO

Girolamo Cardano (1501–76), a mediocre Italian mathematician with exceptional drive and greed—which he apparently bequeathed to his progeny—was the illegitimate son of a prominent Milanese lawyer named

Fazio Cardano, who taught geometry at the University of Pavia. When Fazio was in his fifties, he had an affair with a woman in her thirties named Chiara Micheria, who had become widowed and was forced to care for three young children alone. She became pregnant by Cardano, but before she gave birth, a plague ravaged Milan and she traveled to Pavia, where her lover was teaching, leaving her children behind. Girolamo Cardano was born in Pavia, and when he and his mother returned to Milan, she was horrified to find that all three of her older children had died from the plague. She raised the boy alone, and only late in life did Fazio Cardano marry her.

As a young man, Girolamo worked as his father's legal assistant, but he wanted a lot more out of life, so his father taught him mathematics and enabled him to enroll at the University of Pavia. After considering a career in mathematics, he decided to study medicine, which he thought would be more financially rewarding. When the war with the French broke out, the university closed its doors and Cardano moved to Padua to continue his education. He joined the faculty and even tried to get elected rector of the university. Despite his lack of popularity, which was probably due to his unpleasant disposition, he won the election by one vote. But he did not hold the post for long.

Fazio died that year, and when Girolamo got hold of his inheritance, he became a compulsive and aggressive gambler. At times he was even violent. It was said that his understanding of probability helped him win large sums of money; but when he lost, he would get angry. On one occasion he slashed the face of an opponent with a knife he kept in his pocket whenever he went gambling. This incident brought him infamy and cost him his life's ambition: membership in the College of Physicians in Milan.

Lacking this necessary professional recognition, Cardano had a hard time making a living as a physician and fell into poverty. Forced to move to the small village of Sacco, outside Padua, he married Lucia Bandarini and had three sons with her. They struggled financially and were forced to move to another village and then to Milan, where Cardano was able to teach mathematics and practice medicine on the side.

In Milan, Cardano managed to cure several patients from serious illnesses that other physicians could not treat successfully. This brought him new renown and financial success. He was even invited to Scotland to treat the archbishop of St. Andrews, who was suffering from what we now know is asthma. The archbishop recovered, and Cardano was paid two thousand crowns. Now a successful physician, Cardano again turned his attention to mathematics. His ambition was to make his mark on mathematics history. One day he heard about the great mathematical achievements of the man called the Stammerer. In flowery, flattering language, he wrote Tartaglia a letter, telling him he was publishing a book about equations and needed Tartaglia's help. Moreover, Cardano promised to credit Tartaglia and to enhance his reputation, which would result in better academic and consulting jobs, if only Tartaglia would reveal his secret. Tartaglia refused. Over several years Cardano continued to beseech Tartaglia, always promising better future prospects in exchange for his insight into solving cubic equations. In the meantime, Tartaglia's economic fortunes began to decline as the novelty of his discovery wore off.

In 1539, while Tartaglia was tutoring private students in Venice, a letter from Cardano arrived in which the latter offered to introduce him to the head of the military in Milan—a powerful man who might offer Tartaglia a lucrative appointment as a consultant on fortifications—if only Tartaglia would pay him a visit. Tartaglia finally took the bait. When he arrived at Cardano's house, he was disappointed to find that the promised military leader was not there. Tartaglia was angry, but before he turned to leave, he allowed Cardano to offer him a drink, which led to another . . . and another. Late that night, he gave up his secret.

Cardano's book was published in 1545 and marked what many historians consider the beginning of modern mathematics. Tartaglia, who was credited in the book, was angry, and throughout the rest of his life tried unsuccessfully to stop later editions of Cardano's book, the *Ars Magna* (Great Art), from being published.

Cardano extended the results he conned Tartaglia into providing him, all of which were revealed in the book. Though he thanked Tartaglia three times in *Ars Magna*, Tartaglia was unappeased. The year following the publication of *Ars Magna*, Tartaglia published his own book, in which he reported Cardano's promise and how it had been broken, as well as revealing details of the rest of their conversations. Nonetheless, Cardano's book enjoyed wide success, and he received much of the credit for solving the cubic equation.

After his response to *Ars Magna*, Tartaglia published other books as well, including a 1543 work based on a translation of Archimedes that

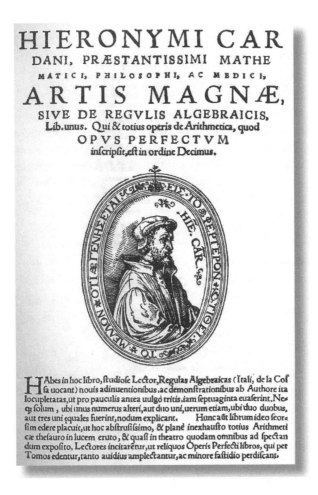

HIERONYMI CAR
DANI, PRÆSTANTISSIMI MATHE
MATICI, PHILOSOPHI, AC MEDICI,
ARTIS MAGNÆ,
SIVE DE REGVLIS ALGEBRAICIS,
Lib. unus. Qui & totius operis de Arithmetica, quod
OPVS PERFECTVM
inſcripſit,eſt in ordine Decimus.

HAbes in hoc libro, ſtudioſe Lector, Regulas Algebraicas (Itali, de la Coſ
ſa uocant) nouis adinuentionibus, ac demonſtrationibus ab Authore ita
locupletatas, ut pro pauculis antea uulgò tritis, iam ſeptuaginta euaſerint. Ne=
q; ſolum, ubi unus numerus alteri, aut duo uni, uerum etiam, ubi duo duobus,
aut tres uni æquales fuerint, nodum explicant.   Hunc aũt librum ideo ſeor=
ſim edere placuit, ut hoc abſtruſiſsimo, & planè inexhauſto totius Arithmeti
cæ theſauro in lucem eruto, & quaſi in theatro quodam omnibus ad ſpectan
dum expoſito, Lectores incitarẽtur, ut reliquos Operis Perfecti libros, qui per
Tomos edentur, tanto auidius amplectantur, ac minore faſtidio perdiſcant.

Girolamo Cardano's *Ars Magna*, the title page of which appears here, was
first published in 1545 under the title *Artis Magnae*.

may have been done by Flemish monk Willem van Moerbeke, whom
Tartaglia didn't credit. Also, in *Quesiti et Inventioni Diverse* (1546) he
presented as his own the law of the inclined plane, which had actually
been derived by Jordanus Nemorarius.[3] Given all these developments, it
has been suggested that perhaps Tartaglia himself learned the formula
for solving cubic equations from another source. Ultimately, Scipione del
Ferro deserves most of the credit for this immense advance in algebra.

In 1540 an eighteen-year-old Italian mathematician named Lodovico Ferrari (1522–60) managed to solve the quartic (fourth-power) equation. *Ars Magna* also included Ferrari's solution method for quartic equations. When Tartaglia attacked Cardano and his book, Ferrari—who was a disciple of Cardano—defended his teacher and claimed that he had been in Cardano's house during the fateful night in 1539 when Tartaglia revealed his "secret." Ferrari argued that it was no secret at all—rather, the method was freely revealed by its possessor. Following this move by Ferrari, Tartaglia challenged him on August 10, 1548, to a contest of solving equations. Ferrari accepted the challenge and later beat Tartaglia at his own game in a public contest in Venice.[4]

What Cardano was able to do in his book was to show how a single root, or solution, can be obtained for a cubic equation satisfying the conditions under study—i.e., lacking a square term and having positive coefficients. In actuality, Tartaglia, Cardano, and others placed terms on one side of the equation or another so that they would always be positive. Whereas $x^3 + ax = b$ and $x^3 = -ax + b$ were treated as different equations, today we know that these are the same equation, except that a term in one form is negative, while in the other it is positive. Mathematicians at that time did not consider negative numbers "real," so Cardano's method only resulted in one root of the equation. We now know that a cubic equation has as many as three roots. The final solution method, leading to the three possible roots, was provided by the great Swiss mathematician Leonhard Euler (1707–83) in an article published in 1732.[5]

In 1557 Tartaglia died in Venice an angry man. Cardano did not enjoy a pleasant old age, either. His eldest son, Giambatista, had married Brandonia di Seroni, a woman from a poor family whose intention seems to have been to extort money from their daughter's wealthy father-in-law. After many failed ploys, Brandonia taunted her husband, saying that he was not the father of her children, and eventually he poisoned her. Giambatista was tortured in jail before his execution in 1560. Another son, Aldo, was a compulsive gambler, like his father, and lost so much money that, at one point, he broke into his father's house and stole most of his money. Cardano made many enemies, which made it hard for him to stay employed. Late in life he moved to Rome and died of an apparent suicide in 1576.

IMAGINARY NUMBERS

While it still evidences the struggle that sixteenth-century mathematicians had with understanding the role and meaning of negative numbers in mathematics, Cardano's seminal book is also the origin of a much more abstract concept: complex numbers. These numbers are combinations of real numbers and multiples of the square root of -1, which are called "imaginary numbers." In many ways—to the mathematician, to the physicist, and to many an engineer—these imaginary numbers are quite real. We denote the base of these numbers—the square root of -1—with i. The term "imaginary," in fact, was first proposed by Cardano.

Toward the end of his book, Cardano dealt with a problem that involved the search for two numbers whose sum was 10 and whose product was 40: the solutions of $x^2 - 10x + 40 = 0$. He realized that no two real numbers can be found to satisfy these requirements. He then proposed what he called a sophisticated approach, in which he said that he could *imagine* the number $\sqrt{-15}$. If that number could be imagined, he said, then the equation and its requirements could be satisfied by noting that the two numbers he was after were $5 + \sqrt{-15}$ and $5 - \sqrt{-15}$:

$$(5 + \sqrt{-15})(5 - \sqrt{-15}) = 25 + 5\sqrt{-15} - 5\sqrt{-15} - (-15) = 40; (5 + \sqrt{-15}) + (5 - \sqrt{-15}) = 10.$$

Thus, imagining a number that satisfies some relations allows us to solve equations that may otherwise be unsolvable. From this one passage in Cardano's *Ars Magna*, imaginary numbers—and hence the whole field of complex numbers—were born.

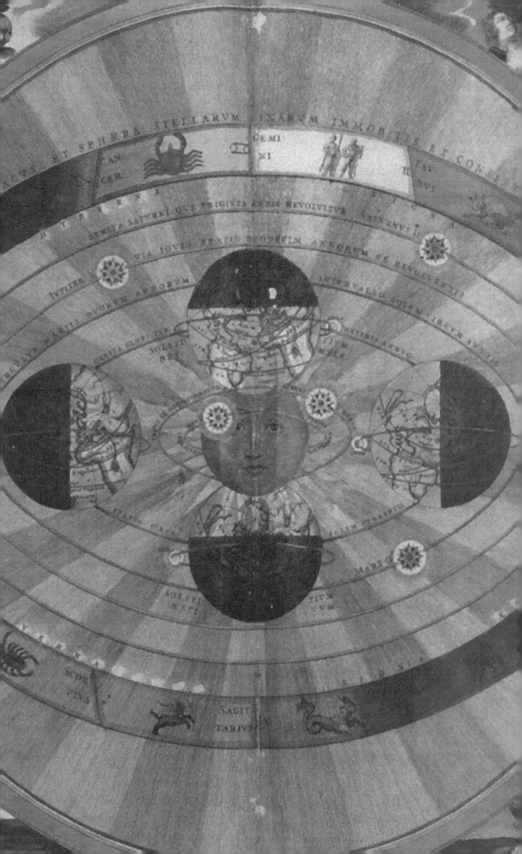

S E V E N

———— •◆• ————

HERESY

s research in mathematics was growing in the sixteenth and seventeenth centuries in Europe, mathematicians were better able to address problems of physics—a science born through extensive efforts in applying mathematics to the physical world—as well as astronomy. This inevitably led to the decline and eventual demise of the millennia-old belief that the earth was the center of the universe, bringing criticism from the Catholic Church, which supported this view as consistent with scripture. Some European mathematicians and scholars ran afoul of the Church because of their scientific conclusions about the position of our planet within the universe. As mathematics developed and sometimes became associated with mysticism,

This plate from cartographer Andreas Cellarius's 1660 tome *Harmonia Cosmographica* outlines the heliocentric (sun-centered) Copernican system.

the black arts, alchemy, astrology, and Rosicrucianism, some mathematicians were viewed as heretics. Mathematics had become a dangerous endeavor.

FRANÇOIS VIÈTE

François Viète (1540–1603), also known by his Latin name, Franciscus Vieta, was born in Fontenay-le-Comte in western France, where he studied in a Franciscan school. His father was an attorney, and his mother also had jurists in her family, so a career in law seemed the natural course to follow.

Upon earning his law degree in the nearby city of Poitiers, Viète returned to his native town in 1560 and started practicing law. He was involved in a few interesting cases early on, one of which concerned the financial affairs of Mary, Queen of Scots. Four years later he took an opportunity that interested him more than practicing law: he became the tutor of Catherine of Parthenay, an eleven-year-old child of the local Protestant nobility. Viète taught her mathematics and astronomy, as well as other subjects. When the girl's father died sometime later, she and her mother moved to La Rochelle on the Atlantic coast, and Viète came with them. There he met many influential members of the nobility, including Henri of Navarre, who later became King Henri IV of France.

In 1570, when Catherine was an adult, Viète moved to Paris, although he remained in touch with her and visited often when he was in western France. He practiced law in Paris and studied mathematics and astronomy on his own. Like the famous Pierre de Fermat, who came after him, Viète was a jurist and avocational mathematician whose work produced important results. He often contemplated mathematical problems while leaning his elbow on a desk or windowsill during his leisure time, and he eventually published a book with the abbreviated title *Mathematical Canon*.

Viète also became a councilor to the Parlement of Brittany. At the time, France was a monarchy, not a democracy, so the Parlement—akin to a parliament—was a body of deputies who advised and served the king. This was the time of the infamous wars of religion in France, and Viète was caught up in the turmoil. He incurred the wrath of the Catholic League when he represented Protestant interests.

FR. VIETE
né en 1540. mort en 1603.

François Viète was an attorney and French government official who
practiced mathematics as an avocation. He was a fervent advocate of
the decimal system, and was the first to use letters to denote unknown
quantities in algebra.

Viète became a member of King Henri IV's council, which gave him
the opportunity to view intercepted Spanish communications. It was in
this capacity that his mathematical skills became most evident, as Viète
had an uncanny ability to decipher coded messages. He was so good at
breaking codes, helping his country gain advantage in its international
conflicts, that the Spanish, upon learning that their secret codes had been
broken by one man, claimed that Viète was in league with the devil.[1]

Viète made many contributions to all the areas of mathematics known in his time, but one of his key contributions was pushing for the adoption of decimal fractions. The cumbersome Babylonian base-60 system was entrenched throughout the world, from India to Arabia and Europe. Interestingly, it was none other than Viète's insistence on the abandonment of numerals based on 60 that finally impelled Europe to adopt the full decimal system we know today.

Viète also did important work in algebra. He provided a new way of solving the cubic equation that involved rewriting the relationship between the variables. He introduced concise notation in algebra, using letters to denote unknowns, and thus transformed the "word-based" algebra of the Italian cossists and their Arab predecessors into the modern symbolic language we use in algebra today. In fact, Viète's notation and methods are said to have catapulted algebra into a new phase in its development.

Viète was especially adept at using trigonometry in algebraic problems. In 1593 a Flemish mathematician named Adriaan van Roomen issued a challenge to mathematicians to solve an equation of the forty-fifth degree: $x^{45} - 45x^{43} + 945x^{41} - \ldots - 3795x^3 + 45x = C$, where C is a number. The Low Countries' ambassador to France flatly declared to King Henri IV that no French mathematician could answer this problem, so the king called on Viète to defend France's reputation. Viète looked at this problem, leaning on his elbow, and immediately recognized that the equation, in fact, stated the trigonometric relationship between $\sin(x)$ and $\sin(^x\!/_{45})$ algebraically. Once he understood the connection with trigonometry, the solution was immediate. He achieved great fame as the mathematician who defended his country's honor, and his use of trigonometry in algebra increased the scope and applicability of the study of trigonometry. When told of Viète's feat, van Roomen is said to have saddled his horse and set off for Fontenay-le-Comte, where Viète was at the time. He stayed with him for weeks, and the two became close friends. In 1602 Viète left the service of the king, receiving a compensation of twenty thousand écu. These were found by his bedside when he died in 1603.

JOHN NAPIER

Trigonometry was of great interest to another mathematician, one who lived far from Italy and France. John Napier (1550–1617), Baron of Merchiston, was a Scottish nobleman chiefly engaged in running his vast estate. He had other interests, however. One of them was the book of Revelation. In a commentary on the book, he argued that the pope was the Antichrist.

Napier was an astrologer and necromancer—he delved into the practice of conjuring the spirits of the dead. He was also interested in alchemy and, in his house, performed strange experiments with flasks of boiling liquids. He was said to have always traveled with a black spider he kept in a little box, and he had a black rooster he used as a "familiar spirit" in magical endeavors. For example, he would force each of his servants to sit in a room alone with the rooster and stroke the bird. This trick allowed Napier to determine which of his servants had stolen from him. In actuality, he is said to have covered the rooster with soot and determined the guilt of a servant by seeing which one had clean hands afterward, with the assumption that the guilty person—afraid of being caught—would only pretend to stroke the rooster.

The Scottish astrologer John Napier introduced the concept of logarithms—a way of lowering the levels of calculations—in 1614.

In 1594 Napier began to think about an idea that would change the world of mathematics: how to simplify arithmetic computations by changing multiplication into addition. He was expanding upon the idea that addition of powers in arithmetic is equivalent to multiplication when the base is the same (e.g., $5^{(2+3)} = 5^2 \times 5^3$). Napier hoped to apply a variant of this rule to other computations, but he couldn't find a good way of doing it until his friend John Craig, personal physician to King James VI of Scotland, told him a story. Craig had been with the king when he sailed in 1590 to Denmark to meet his bride-to-be, Anne of Denmark. Just before arriving in Copenhagen, heavy storms in Oresund Strait forced them to come ashore on Hven Island, on which the famous Danish astronomer Tycho Brahe had situated his observatory.

Brahe entertained the king and his entourage, and Dr. Craig learned that, in his calculations of astronomical observations, Brahe was making use of the method of prosthaphaeresis—using trigonometric identities to reduce multiplication to addition (and division to subtraction)—which had been discovered by Ibn Yunus five centuries earlier. This story encouraged Napier to redouble the efforts he had been making to find an even more efficient way of lowering the levels of computations and, in 1614, after finishing the Herculean job of computing logarithms, he published his book, *Mirifici Logarithmorum Canonis Descriptio* (A Description of the Marvelous Rule of Logarithms).[2]

For three hundred and fifty years—until the middle of the twentieth century, when calculators became readily and inexpensively available—Napier's logarithms ruled the world of calculation. The slide rule, which was the mechanical forerunner in the West of the electronic calculator, was based on Napier's idea of the logarithm. The slide rule was marked with a logarithmic scale of numbers, and it allowed a person to perform multiplication through the actual physical "addition" of two sections of numbers. This was accomplished by sliding the central strip of the ruler back and forth and reading the answer on the fixed strips on the top and bottom.

JOHANNES KEPLER

Tycho Brahe was a very colorful figure. Having lost his nose in a duel while he was a student, he wore a prosthetic nose made of metal. He was a nobleman close to the king of Denmark, who had given him the island of Hven on which to build his observatory. There, Brahe carried out the most extensive observations of the sky ever made by one person, as far as we know. His work led him to a theory that fell somewhere between the Copernican model of the universe and the earth-centered model that preceded it. Drawing upon his observations of a supernova, he showed that the stars were far above both our atmosphere and the moon, belonging to a higher "sphere." This and other observations allowed him to overthrow the Ptolemaic model, which had reigned for almost a millennium and a half. After a quarrel with the king, Brahe left for Prague, where he was given facilities to continue his work under the Bohemian king Rudolph II. The brilliant German mathematician Johannes Kepler became his assistant.

Johannes Kepler (1571–1630) was born prematurely in Weil der Stadt, Germany, to an innkeeper's daughter and a mercenary. His father, Heinrich Kepler, abandoned the family when Johannes was just five years old and is believed to have perished in the Thirty Years' War. At an early age Johannes impressed lodgers at his grandfather's inn with his mathematical ability, and after witnessing the Great Comet of 1577 and a lunar eclipse in 1580, he developed a lifelong interest in astronomy. Childhood smallpox left Johannes with weak vision and crippled hands, but in 1589 he enrolled at the University of Tübingen and quickly gained a reputation as a skilled mathematician and astrologer. Despite his desire to enter the ministry, Johannes was recommended for a professorship at the University of Graz, which he accepted at the tender age of twenty-three.

During his tenure in Graz, Kepler studied the conic sections identified in ancient Greece because he was interested in mirror images of various shapes. In 1595 he married Barbara Müller, a twenty-three-year-old widow (twice over) with whom he had three children. Five years later he was invited by Brahe to assist in calculating planetary orbits at a new observatory he was constructing outside Prague. The study of ellipses and the circle, in conjunction with an analysis of his employer's vast set of astronomical

Johannes Kepler, depicted in this 1610 portrait, served as Tycho Brahe's assistant and mathematically deduced his famous laws of planetary motion.

observations of the planets, allowed Kepler to make one of the most important scientific deductions in history—the laws of planetary motion:

1. Each planet moves around the sun in an elliptical orbit that has the sun at one of the two foci of the ellipse.

2. The radius line that connects the planet to the sun sweeps equal areas in equal time.

3. The square of a period of a planet is proportional to the cube of its orbit's semimajor axis.

Kepler viewed the areas swept by the radius as comprised of an infinite number of tiny triangles, each with an infinitesimally small area. By "summing up" these areas mathematically, he was using a forerunner of the idea of integral calculus, which would be further developed by Descartes and later formalized by Leibniz and Newton.

After Tycho Brahe's unexpected death in 1601, Kepler succeeded him as Imperial Mathematician to Holy Roman Emperor Rudolph II (the former king), a prestigious position he held until the emperor's abdication eleven years later. In 1612 his wife and son contracted illnesses and died. That year, Kepler found an effective way to estimate the volumes of casks of wine. In carrying out this estimating procedure, Kepler went far beyond the work of Archimedes on volumes and, as he had done with areas, built an argument about the summation of an infinite number of elements of volume, each infinitesimally small—another close forerunner of integral calculus.

Kepler is also famously known for incorporating Plato's five solids into a model of the universe. There were only five known planets at the time, including Earth, so Kepler concluded that there were five "separators" among their orbits around the sun. Thinking that perhaps he had found a cosmic meaning for the five mathematical elements of Plato—as the Greeks had done two millennia earlier by ascribing the elements of nature (earth, wind, water, fire, and quintessence) to these solids—he

Kepler used the idea of Platonic solids (see page 24) to create this diagram of the solar system, which appeared in his 1596 book *Mysterium Cosmographicum*.

placed each planetary orbit on a sphere that inscribed one solid and was inscribed by another. It was a valiant attempt to impose mathematical structure on heavenly bodies, but—of course—it was false. (And it contradicted Kepler's own discovery of elliptical orbits.)

By the end of the seventeenth century, not only did Newton and Leibniz incorporate the methods used by Kepler to find areas and volumes into integral calculus, but Newton's laws of gravitation evidenced the overarching principles from which Kepler's laws could be mathematically derived—using the same calculus.

In 1617 Kepler's mother, Katharina, an herbalist who had encouraged his interest in celestial events, was accused of witchcraft. Ursula Reingold, who was involved in a financial dispute with Kepler's brother, claimed that she had contracted an illness from an "evil brew" Katharina had given her. Johannes went to great lengths to defend his mother in court, and after fourteen months in jail, she was released on technical grounds relating to the use of torture to extract "evidence." (Not every accused witch was so lucky; of the fifteen women accused of witchcraft under the reign of Lutherus Einhorn, an overseer in Leonberg, eight were executed.) Kepler himself wrote horoscopes and indulged in mysticism, which, in combination with his advocacy for the Copernican heliocentric system, frequently landed him in political trouble. In 1625 measures enacted under the Catholic Counter-Reformation placed almost his entire body of work under seal. He died on a visit to Regensburg in 1630, where he had traveled to collect money owed him for one of his books.

GALILEO GALILEI

A contemporary of Kepler's, Galileo Galilei (1564–1642) was born in Pisa, Italy, to a famous lutenist and musical composer. His contributions to physics, astronomy, and cosmology are immense. In 1609–10 Galileo used the recently invented telescope to look at the night sky and discovered the moons of Jupiter—now referred to as the Galilean satellites. This discovery lent much support to the developing arguments against Aristotelian philosophy and the Ptolemaic cosmological system, which placed Earth at the center of creation. Galileo also performed many experiments with falling objects—perhaps even dropping them from the top of the

Leaning Tower of Pisa, as has been conjectured—which led him to discover that an object's time of descent is independent of its mass. Galileo made important mathematical discoveries, too. By studying trajectories in the air, he was able to show that the path of a projectile in the air is a parabola. He also studied the cycloid—the curve traced by a point on the rim of a wheel as the wheel moves along a horizontal path.

Although he never married, Galileo fathered three children out of wedlock with a young Venetian woman named Marina Gamba. His two daughters were thus considered unmarriageable and were sent to a convent in Arcetri. Galileo's eldest daughter took the name Sister Maria Celeste and remained devoted to her father until her death in 1634.

In pure mathematics, Galileo made a key discovery about infinity, referred to as Galileo's paradox. He found that even though only some whole numbers are squares, and hence the set of whole numbers should be larger than the set of squares, he could set up a one-to-one correspondence between the set of whole numbers and the set of squares of whole numbers. Simply by pairing all positive integers with their corresponding squares—e.g., $1 \to 1$, $2 \to 4$, $3 \to 9$, $4 \to 16$, and so on to infinity—both sets could be "counted" against each other. This led him to the conclusion that there are "as many squares as there are numbers."[3] Later on, we will see why this discovery was so important in the study of infinity.

THE INFINITE HOTEL

In the twentieth century the German mathematician David Hilbert gave an entertaining example of the fact that an infinite set can be "counted" against a subset of itself, and thus both infinite sets can be "equal" in size despite the fact that one set contains the other. The example is called Hilbert's Hotel, or the Infinite Hotel, and it proceeds thus: A person arrives late at night at the Infinite Hotel but is turned away.

"Sorry, we are completely full."

The visitor protests, "But the brochure says that this is an infinite hotel—you have infinitely many rooms."

"Yes," responds the receptionist. "But all our infinitely many rooms are full."

"Okay," says the visitor, "this is what you should do: move the guest from room number 1 to room number 2; the one in room number 2 to room number 3; and so on. Since you have infinitely many rooms, you can do this for all your guests. Then room number 1 becomes available for me."

Here, we have the one-to-one correspondence 1→2, 2→3, 3→4 . . . so the set of all integers and the set of all integers save the number 1 are equivalent to each other, and hence include the same number of elements in each (though it is infinite). Indeed, this trick shows—as Galileo had understood in the 1600s—that an infinite set can still be put into a one-to-one correspondence with a proper subset of itself.

As portrayed in this 1857 painting by Cristiano Banti, the great Italian mathematician Galileo Galilei faced trial by the Roman Inquisition because of his view that Earth revolved around the sun.

Galileo's contributions to philosophy, mathematics, physics, and astronomy are far too numerous to list here. Despite his immense influence, however, his scientific discoveries led him to question scripture, and his publications incurred the ire of the Inquisition. In 1633 he was tried for heresy in Rome and condemned to house arrest for the rest of his life. He died in 1642, a broken man.

As the story of mathematics continues, we move from the Father of Modern Science to the Father of Modern Philosophy. René Descartes and his contemporaries extended mathematical principles to philosophy, and the seventeenth century saw the rise of rationalism, which anchored reality in geometry and the powers of deduction.

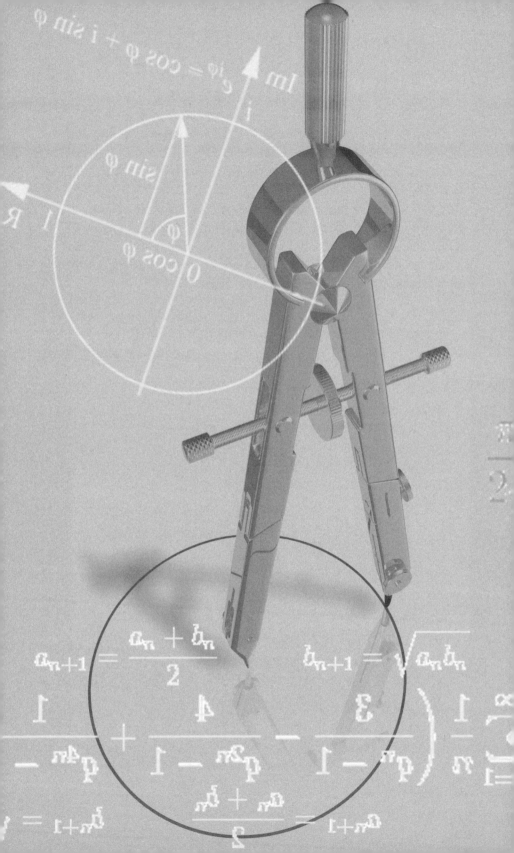

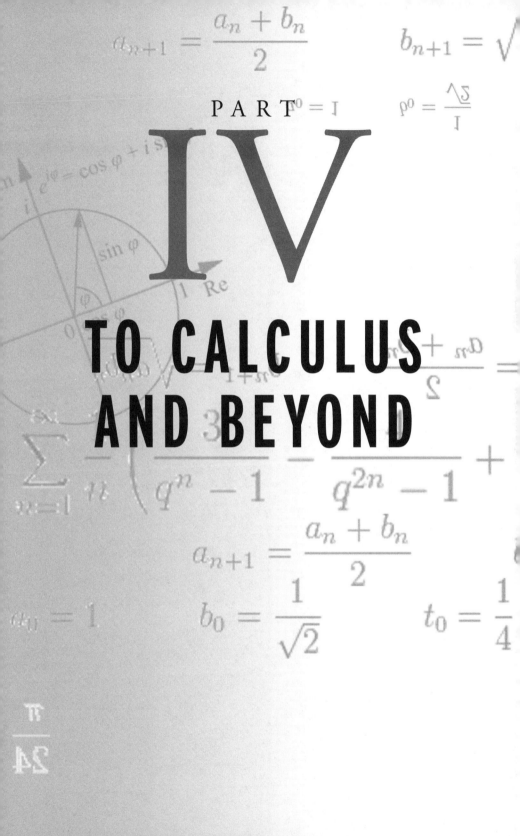

IV

TO CALCULUS
AND BEYOND

THE GENTLEMAN SOLDIER

n the century after Cardano's life, the great French philosopher and mathematician René Descartes (1596–1650) read the *Ars Magna* but wasn't very impressed with it. He was much more enchanted with Galileo's work; he felt a kinship with the Italian thinker—so much so that he worried he might encounter a fate similar to that of Galileo. Like Galileo, Descartes was engaged in research that was leading him in the direction of the Copernican system so abhorred by the Church.

On his own, this exceptional genius had generalized a number of Greek and Arab discoveries in mathematics and was on his way to unifying geometry with algebra. In a

René Descartes—famous for his dictum "Cogito ergo sum"—was hugely influential in mathematics and philosophy. His system of numerical coordinates served as the first codified link between algebra and geometry.

moment of almost careless doodling, he was able to solve one particular kind of *fourth-order* equation (although it should be noted that the Italian algebraists of the previous century could solve some such equations as well). René Descartes was almost too great a genius to even worry about solutions to equations; he had grander ambitions.

Nothing about René Descartes—the great French mathematician, philosopher, physicist, and natural scientist—is what meets the eye. In one of the greatest ironies in the history of ideas, the man who gave us the strict, perfectly logical Cartesian rules seems to have defied every rule of reason in his personal life.

RENÉ DESCARTES

René Descartes was born to a wealthy aristocratic family on March 31, 1596, in the town of La Haye (now named Descartes, in his honor), in the region of Touraine, France. His father, Joachim Descartes, was a councilor to the king of France, working on legal and legislative matters in the Parlement of Brittany in the city of Rennes, roughly one hundred miles to the west. The Descartes family was a major landowner in the agriculturally rich area of Châtellerault, and its members could easily live off the rents paid by their tenants without having to worry much about money.

In those days people were born at home, but the house in which Descartes was born was not his family home. Just as she was about to give birth, Descartes's mother, Jeanne Brochard, left the family mansion in Châtellerault, in the region of Poitou—some twelve miles to the south—crossed the Creuse River, and gave birth at her mother's house in La Haye. She didn't feel comfortable having the baby in her own house, given that her husband was away.

At the turn of the seventeenth century, the two neighboring regions, Touraine and Poitou, were very different. Whereas Poitou was mostly Protestant, Touraine was predominantly Catholic—as is most of France

Marin Mersenne (1588–1648), a monk, music theorist, and mathematician, served as an intermediary between Descartes and other scholars throughout Europe, including Pascal and Fermat.

today. This accident of birth—being born in a heavily Catholic part of France to a family that, although Catholic, hailed from a Protestant region—affected Descartes's feelings about religion and society throughout his life. One key example of this would be his excessive, almost irrational fear of the Inquisition and what it might do to him if he published scientific writings contrary to Church doctrine; at the same time, he remained almost naively unprepared for the attacks on his views and writings that would come from Protestant theologians after he developed his theories as an adult.

Descartes was a weak child with many minor health problems, so from early childhood he had the privilege of sleeping late while the other schoolchildren were in class, joining them only when he felt ready to face the day. Not having to deal with the nuisances of school routine, the young genius was able to derive mathematical results while lying in bed. As he grew stronger, Descartes learned to fence, and he practiced swordsmanship throughout his adolescence and early adulthood.

Once he recovered from whatever had ailed him as a young boy, Descartes's family sent him to study at the Jesuit College of La Flèche, in the nearby region of Anjou, to the north. At La Flèche, Descartes continued to practice his swordsmanship. His studies were focused primarily on the classics and Greek mathematics—especially geometry. What he loved most were the ancient Greek mathematicians' constructions with a straightedge and compass, at which he excelled.

During his studies he met Marin Mersenne, who was a few years ahead of him at school and equally interested in mathematics and science. Mersenne, who would later become a Minim monk in Paris, became a lifelong friend and confidant of Descartes. Often, while Descartes traveled throughout Europe, Mersenne would be the only person who knew his whereabouts and could reach him by mail. And when Descartes wrote a letter to a mathematician, he usually used Mersenne as an intermediary.

Upon graduation Descartes moved to the big city: Paris. There he lived the exciting and carefree life of a young, rich dilettante. He gambled and drank and caroused with beautiful women—especially young women with pretty eyes. He also pursued mathematics, but it was difficult to find the solitude he needed in order to study. Often he would hide in his room when working on a theorem or problem, but his many friends would come and beckon him to join them in cafés and bars or on the streets. Hiding out became a tough task indeed, and Descartes looked for ingenious ways of avoiding disruption. But his friends were good at finding him.

At the time, Saint-Germain-des-Prés was an area at the edge of Paris by the meadows (*prés* in French) where young men would go to duel in secret. Dueling, which was a common practice in Europe among the aristocracy, had been recently made illegal by royal decree, so young men were forced to seek out deserted areas in the fields in order to indulge in it. Although Saint-Germain-des-Prés is now one of the central locations in Paris, with its famous church and trendy cafés, in Descartes's time it was a forgotten backwater just outside the city walls. Descartes told no one that he had taken a room in this remote area—except for his valet, who followed him on his travels throughout the continent. One of his friends discovered his whereabouts, however, when he followed Descartes's valet through the streets when the latter was buying food for his master. Thus the valet inadvertently led the young man right to his master's hideout.

DESCARTES WAS A RESTLESS SORT who hated to settle in one place. In 1618 the twenty-two-year-old decided he'd had enough of Paris and its attractions; he now wanted to become a soldier. Having had much practice with his swordsmanship both in Saint-Germain and in school, he wanted to put his skills to the test, so he traveled with his valet to Breda in Holland and joined the army of Maurice of Nassau, Prince of Orange. Maurice was a Protestant ruler who was planning a battle against Catholic armies during the Thirty Years' War. Descartes accepted a gold coin as token compensation for his entire future service to the prince, and, ever accompanied by his valet, the dandy soldier traveled throughout the region, dressed in green taffeta and carrying a shiny sword. On

November 10, 1618, at the central square of Breda, Descartes had an unexpected revelation that would change his life. Many people were crowded around a tree in the square, to which a piece of paper had been posted. Descartes could not understand Dutch, so he asked the person nearest to him in Latin—the lingua franca of intellectuals throughout Europe at the time—what the writing was. The man Descartes happened to ask was Isaac Beeckman, a physician and would-be mathematician who fancied himself much smarter than some random French soldier.

"It's a mathematical problem," he answered.

"I can see that!" snapped Descartes. "But what does it say? I don't understand Dutch."

Beeckman translated it for him and then added arrogantly, "And I suppose you'll give me the answer once you've solved it?"

"Of course I will!" answered Descartes, looking him intently in the eye. He asked for his address, and Beeckman explained that he was from out of town and staying with his uncle to help him slaughter his pigs. He was also looking for a wife, he added. Descartes took Beeckman's uncle's address and departed.

The next morning, very early, Descartes knocked at Beeckman's uncle's door and gave the stunned Dutchman the solution to the puzzle from the poster, which had stumped not only all passersby but also the mathematics professors at the local university. We don't know what the problem was, but thanks to Beeckman's journal, which was unexpectedly discovered in a Dutch library in 1905, we know that it was a geometrical one, perhaps based on ancient Greek theorems.

From that triumphant moment, Descartes knew he was a gifted mathematician, but he was also determined to never again require translations from Dutch. During long stays in Prince Maurice's camp, Descartes taught himself the language of his fellow soldiers. Meanwhile, Beeckman and Descartes enjoyed a long correspondence, mostly about mathematical problems and ideas.

Descartes had heard that in Bohemia a Protestant king, Frederick II—later to be called the Winter King, as his rule would last only one season—was besieged by Catholic forces. Prince Maurice's army could not travel there, however, since the Protestant leader's predecessor had signed a treaty with Spain in 1609 that prevented the Dutch army from waging war on anyone for twelve years. As a volunteer soldier, Descartes was eager to see

This depiction of the Temple of the Rosy Cross was designed by alchemist Daniel Mögling (a.k.a. Theophilus Schweighardt) in 1618. The temple has wheels to signify that the abode of the Rosicrucians is "nowhere but everywhere"; the hand of God holds the rope that guides it from above.

action, so one day he gathered his valet and headed southeast toward Bohemia (the modern-day Czech Republic). On the way, he stopped in Frankfurt, arriving there just in time to witness the elaborate ceremony of the coronation of the Holy Roman Emperor. Then he continued south and arrived in the German city of Ulm, where he met a mystic mathematician.

Johann Faulhaber was a rumored member of the Order of the Rosy Cross. The Rosicrucians, as they were called, were a secret society of scholars concerned with promoting science—which included alchemy—and mathematics. They wanted to cure all the ill people in the world and, being opposed to institutionalized knowledge, they supposedly favored Protestant causes that challenged the authority of the Church. But no one had ever met a Rosicrucian: they were known as the Invisibles, since they were always in hiding. Because no one ever saw them, one could say anything about them, including that they had never existed. Descartes's first biographer, Adrien Baillet, who wrote about the philosopher-mathematician's life only forty years after his death, claimed that Descartes was involved with the Rosicrucians.

Faulhaber received Descartes warmly and took him into his library. He presented him with some algebraic problems that Descartes solved admirably well and very quickly. Faulhaber laughed in delight and gave the young soldier more problems—all of which he solved with ease and elegance. The two men became close friends who exchanged mathematical ideas throughout life, and Descartes went as far as to adopt Faulhaber's mystical

♃

The symbol for Jupiter—which Descartes used in his algebraic calculations—is also the alchemical symbol for tin.

The Battle of White Mountain was fought on November 8, 1620, near the Bohemian capital of Prague, which was then in the grip of rebel forces. The battle, in which Descartes fought for Maximilian in the combined forces under Ferdinand II, is memorialized in this painting by the Flemish artist Pieter Snayers.

notation. For example, he used the sign of Jupiter in his algebraic manipulations. This sign originates in astrology and alchemy.

Descartes continued his travels and—in a reversal, having served in a Protestant army—now joined the Catholic army of Duke Maximilian of Bavaria in the battle for Prague, which unseated Frederick. The Winter King, his wife, and their young daughter Elisabeth fled Prague in the middle of the night as Descartes and his fellow soldiers stormed the city. Thereafter, only a Habsburg would rule over Bohemia. Descartes and Elisabeth, "brushing" against each other that night, as it were, would meet in Holland years later and become close.

During the long winter of encampment outside Prague in 1619–20, Descartes lived in a *pôele* (French for "oven"), as he described it. It was a small hut containing a woodstove for cooking and heat. On the night of November 10, 1619, which was coincidentally the anniversary of his encounter with Beeckman in Breda, Descartes, sleeping in his "oven" by the Danube, experienced three intense dreams that apparently had a very strong influence on his life and quest for truth. The next day, he wrote in his notebook about a key mathematical discovery he claimed to have made that night. He did not specify what the discovery was or what area of mathematics it applied to, and the nature of this early breakthrough by the young Descartes remains a mystery.

After the battle for Prague, Descartes continued on his travels. One day, returning to the region of Poitou to sell some of his land in order to finance more travel, Descartes stopped by a roadside inn at the crossroads of two major routes near Orleans, south of Paris. Just outside the inn, he recognized a young woman with whom he had had an infatuation some years earlier. She recognized him, too, and rushed to meet him—and it was as if the intervening years had done nothing to cool their passion for each other. Suddenly her current suitor leaped toward Descartes and challenged him to a duel. Apparently, the man had no idea whom he was dealing with: an expert swordsman.

Needless to say, Descartes defeated the suitor easily. With one quick parry-and-attack move, Descartes made the man's sword fly up in the air and then quickly put the point of his sword to his opponent's neck. He said, "The lady has beautiful eyes, and for that I will spare you your life." And with that, he straightened up, gathered his valet and their horses, and galloped away, leaving his stunned former love and her suitor in the dust.[1]

In July 1621 Descartes traveled from Germany to Hungary, Moravia, and Silesia, where he observed the war's devastation on the city of Breslaw. The following autumn, he continued on to Pomerania, by the Polish border, stopping in the city of Stettin. He went on to the Baltic coast, to Brandenburg, the Duchy of Mecklenburg, and then to Holstein. Descartes was determined to visit most of the continent of Europe.

In November 1621 Descartes decided to make one last trip before settling down in one place, choosing to see the Frisian Islands off the northern coast of Germany and Holland. He and his valet hired a Dutch boat with a crew to show them the islands in detail. They had barely reached the open sea when the disreputable crew, eyeing the French gentleman with what was clearly a bagful of money, spoke freely in Dutch and within earshot of Descartes about their plan to rob him and throw both men overboard. Having made a special effort to learn Dutch while serving Prince Maurice, Descartes was fully aware of the unfolding conspiracy, however. Once he had heard enough, he lunged toward them with his sword and cursed them in their own language. Apparently, they were so stunned by this surprise move that "they failed to consider the advantage of their number and surrendered and peacefully brought him and his valet to their destination," as Adrien Baillet described it in 1691.

Meanwhile, Descartes's good friend Marin Mersenne had taken up residence at the Minim monastery in what is now the Place des Vosges in the Marais section of Paris—a medieval district that was not razed to the ground by Baron Haussmann to make way for wide boulevards in the nineteenth century and, therefore, still looks as it did in the time of Descartes. Engaging in an extensive correspondence with the most prominent mathematicians and scientists throughout Europe, Mersenne participated in what would later be called the Republic of Letters. When he died, ten thousand letters were found in his quarters in the abbey. These letters revealed an extensive, if indirect, correspondence between Descartes and several prominent mathematicians of the time, including Fermat, Pascal, Roberval, Desargues, and other mathematicians, with Mersenne as intermediary.

Progressively, Descartes became obsessed with the Inquisition. He was painfully aware of the travails of Galileo and was determined not to encounter a similar fate. In letter after letter to Mersenne, he expressed his fear that if he published a work proving that the earth rotates around the sun, he would be hunted down and imprisoned—or worse. He had written a book called *Le Monde* (*The World*) but refused to publish it because it espoused the heliocentric view. To further protect himself, Descartes would usually post his letters to Mersenne from locales near his residence rather than from wherever he was residing at the time.

MERSENNE PRIMES

Mersenne was a good mathematician in his own right. He was especially interested in prime numbers and looked for ways to identify them more efficiently and accurately than was then possible through use of the ancient Sieve of Eratosthenes. In particular, Mersenne was able to identify a relationship between prime numbers in his formula $M_p = 2^p - 1$, where M_p is potentially a prime number and p is a known prime number (e.g., $2^2 - 1 = 3$, a prime number; $2^3 - 1 = 7$, another prime number). This relationship holds true until $p = 11$, as $2^{11} - 1 = 2,047$, which is the product of 23 and 89, and thus not prime.

Even though it turns out that so-called Mersenne numbers (M_p) are not always prime, the largest prime number found to date is a Mersenne prime ($2^{43,112,609} - 1$). The search for greater and greater prime numbers continues to this day, and thanks to Mersenne, a project called the Great Internet Mersenne Prime Search facilitates the process of identifying prime numbers by inserting the largest known prime number into Mersenne's formula to see if the result is prime. If it is, then the new prime number is inserted into Mersenne's formula as the search continues.

In October 1628 Descartes witnessed the horrific siege of La Rochelle by Louis XIII and his army. Here was a French king waging a religious battle against his own citizens, the Huguenots. Fleeing persecution, these French Protestants found refuge in the city of La Rochelle on the Atlantic coast of Brittany, where they were promptly surrounded by French forces that blocked all attempts by land or sea to resupply the city's inhabitants with food and other necessities. After fourteen harrowing months, more than twenty thousand Protestants had starved to death, and the remaining population of five thousand—which had been reduced to eating rats and leather belts—finally surrendered. According to Baillet, this was the last battle that Descartes ever observed.

ALTHOUGH THE KING of France had given him special privileges based on his status as a prominent mathematician, philosopher, and scientist, Descartes—who was then settled in Paris—saw a danger to himself in living in a predominately Catholic nation. By the end of 1628, he had left for Holland, seeking tranquillity. It did not last.

Upon receiving word of Galileo's 1633 trial by the Inquisition, Descartes's obsession with the Italian scientist's fate intensified, and he became increasingly distraught. Between 1628 and 1649, he changed addresses more than a dozen times and kept withholding publication of his book *Le Monde,* which described the motion of planets around the sun. Also, his

This detail of a painting by an anonymous artist depicts the surrender of the Huguenots at the conclusion of the 1627–28 Siege of La Rochelle, the last battle at which Descartes was present.

longtime friendship with Beeckman fell apart when the latter boasted that he had "taught Descartes all the mathematics he knows."

While renting an apartment at present-day 6 Westermarkt Street, near the West Church in Amsterdam, Descartes fell into a romance with his landlord's servant, Hélène Jans van der Strom. Unlike most servants, Hélène was literate, and the two continued to exchange letters for years after their affair ended. The couple kept their relationship secret, and on July 19, 1635, they welcomed a daughter into the world, whom Descartes named Francine, meaning "little France." Descartes had intended to send her to France so that she could receive a good education, but in 1640, at the age of five, she succumbed to scarlet fever and died. Distraught by their loss, the couple eventually parted ways.

In 1637 Descartes finally published his masterpiece, *The Discourse on the Method*, which contained the essence of his philosophy, including the timeless statement "Cogito ergo sum" (I think, therefore I am). The book's appendix, entitled *La Géométrie*, contained Descartes's breakthrough work on uniting geometry with algebra, which laid the foundation for the ubiquitous Cartesian coordinate system. *Discourse* made Descartes immensely famous all over Europe.

Elisabeth, Princess of Bohemia, was living at that time in exile in The Hague, an administrative center in the west of the country. Two decades his junior, the beautiful princess—who had crossed his path as she fled Prague with her parents—became his student in philosophy and mathematics.

Descartes (far right) and Queen Christina of Sweden (far left) are depicted in this detail from a painting of the philosopher among members of the queen's court. Descartes tutored the monarch daily.

They corresponded for many years on questions of philosophy, and he even sent her derivations of theorems in geometry, which apparently delighted her. Although Descartes had no shortage of friends in high places, enemies began to emerge out of the woodwork. In 1647, while living in the Dutch countryside, he became entangled in one of the most vicious academic disputes in history. Protestant theologians at the University of Utrecht took issue with Cartesian philosophy, which is based on doubt. Descartes defended his views in print and was subsequently accused of libel by Gisbert Voetius, a man who had also accused Descartes of atheism. A court decided against Descartes, and he was forced to write an official apology, taking back what the court believed to be libelous statements against Voetius. The philosopher reluctantly complied, but there was a growing feeling in his heart that there was no longer a place for him in Holland.

The possibility of escape came from an unexpected source: Queen Christina of Sweden had heard about the famous French philosopher living in self-imposed exile in Holland and decided that she wanted him for herself—as a private tutor. She wrote him letters, which flattered him, but he was still not ready to leave the comforts of Holland. The queen—a strong-headed woman, even though she was only twenty at the time—refused to give up. At one point she simply sent Admiral Fleming of the Swedish Royal Fleet to pick up the reluctant philosopher. Descartes relented and came aboard the ship. After settling in Sweden, he gave the queen lessons in philosophy at 5:00 a.m. every day in the unheated library of her palace.

But Descartes had enemies in Sweden as well. Some court toadies didn't like the newcomer's influence on their queen and were determined to put a stop to it. One of them was the queen's doctor.

In early 1650 Descartes fell ill. He had contracted the flu from the French ambassador, and his condition deteriorated. The queen's doctor insisted on bleeding him, which was a standard medical practice in those days. Descartes knew enough about medicine to refuse the treatment, but eventually, when his condition took a sharp turn for the worse, he consented. On February 11, 1650, the great French philosopher and mathematician died— whether from the flu or from poisoning, we do not know. His remains, along with his mathematical and philosophical papers, were brought back to France, and his bones were buried inside the church of Saint-Germain-des-Prés in Paris—in the area in which he loved to roam. Descartes's skull, however, is now part of a tasteless display of anthropological finds in the Musée de l'Homme (Museum of Man) near the Trocadéro in Paris.

THE JURIST FROM TOULOUSE

Pierre de Fermat (1601-65) could not have been more different from his rival Descartes. Whereas Descartes was a soldier and adventurer, Fermat always led an exceptionally quiet existence—in fact, some might have called him boring. Unlike Descartes, who was interested in metaphysics, philosophy, and the applications of mathematics in physics, Fermat was interested almost exclusively in pure mathematics—save for some applied work in optics. In the field of pure mathematics, however, Fermat is sometimes described as the best of his generation.

Pierre de Fermat was born in the French town of Beaumont-de-Lomagne to a family of leather merchants. He studied classics at the University of Toulouse. (Interestingly, there is no evidence that he studied much mathematics there.) At age thirty he was appointed commissioner of requests in Toulouse and married in the same year.

With his wife, Louise de Long—a cousin of his mother's—Fermat had three sons as well as two daughters, who eventually became nuns. In 1631 he was appointed councilor in the Parlement of Toulouse, a position he held until he died in 1665. According to the historian Eric T. Bell, Fermat fulfilled his duties to the king "with dignity, integrity, and great ability for seventeen years," but he spent all his free time studying mathematics.[2]

Fermat studied graphs of equations, developing the kind of connection between algebra and geometry that Descartes explored, and he looked for ways to study rates of change as well as the maxima and minima of functions. Thus Fermat was another mathematician whose work aided the development of calculus. According to the historian Michael Mahoney, almost everything that Fermat did in mathematics—for example, solving complicated equations—was inspired by the work of François Viète.[3]

After reading Diophantus's *Arithmetica*, Fermat became very interested in Diophantine equations, which led to his preoccupation with number theory. It was, in fact, in the margin of a Latin copy of the *Arithmetica* that he wrote his famous Last Theorem.

Fermat's Last Theorem is the statement that the equation $x^n + y^n = z^n$ has no solutions in integers for any power n that is greater than 2. Produced by someone who was technically an amateur mathematician (Fermat, after all, had a "day job"), the theorem has attracted thousands of amateur and professional mathematicians over three and a half centuries to the cause of proving it. Only in the mid-1990s did Andrew Wiles of Princeton University, with the help of another British mathematician, Richard Taylor, present papers proving that the conjecture is, indeed, correct. In 1637, when he proposed the theorem, Fermat wrote on the margin of the book, "I have a marvelous proof of this assertion, but the

Pierre de Fermat, portrayed in this undated engraving, was a "pure" mathematician, interested chiefly in mathematical theories and principles rather than their applications.

margin is too small to contain it." Did he possess a proof? Probably not. When the final proof was obtained by Wiles, he had to use more than two hundred pages of derivations based on mathematics that had been developed centuries after Fermat's time.

Descartes, when exchanging information with Fermat (through Mersenne) about pre-calculus ideas and the union of algebra and geometry, referred to Fermat not by name but as "the jurist from Toulouse," which took the mild-mannered Fermat aback. Each of them vied to be the greatest mathematician of his time.

DESCARTES LEFT AN IMMENSE LEGACY in mathematics and philosophy. His "I think, therefore I am," statement, as well as his work on metaphysics and the mind-body connection, are mainstays of modern philosophical thought. In mathematics he understood enough about rates of change of functions, and about areas, to be considered one of the key theorists who paved the way for the invention of calculus.

While on the road, in the countryside of various European nations, or hiding out in Paris, Descartes successfully managed one of the greatest feats in mathematical history: he wed geometry with algebra. In other words, Descartes was able to show how every geometrical object can be associated with an algebraic equation. For example, a parabola, as every high-school student learns, is associated with a quadratic equation. In setting up this correspondence between geometry and algebra, Descartes gave us the Cartesian coordinate system, which allows us to associate each point in space with a set of numbers that describes its position. A modern piece of technology directly enabled by Descartes's discovery is the Global Positioning System (GPS), which uses the Cartesian coordinate system to identify every point on Earth by its longitude, latitude, and altitude (height above sea level). In addition to this crucial development in mathematics, Descartes made a large number of beautiful mathematical discoveries, but he did not publish his findings until many years later. Other discoveries, found in notebooks he had left by his deathbed in Stockholm, remained unpublished until after his death.

———— •◆• ————

NINE

THE GREATEST RIVALRY

I n 1676, twenty-six years after the death of Descartes in Stockholm, a budding German intellectual living in Paris was determined to read all the works of the late French mathematician—both those that had been published and those that were hitherto unknown to the general public. The young German man managed to find a French gentleman named Claude Clerselier, a relative by marriage of Descartes. Clerselier had also been Descartes's publisher and now jealously guarded all the documents that the great philosopher and mathematician had left behind when he died. It is a miracle that these papers survived: after the French ambassador to Sweden shipped them to Paris

The rivalry between Gottfried Leibniz and Isaac Newton over who had first discovered calculus lasted until Leibniz's death in 1716. In this frontispiece to Voltaire's 1738 *Eléments de la philosophie de Newton*, the light of Newton's wisdom is reflected onto Voltaire from the heavens by his muse and lover Émilie du Châtelet, who translated Newton's work for him.

following Descartes's death, the boat carrying the documents capsized while sailing up the Seine. Amazingly, the box containing these precious writings—most of them unpublished—was found floating on the river, and Clerselier and his servants spent days drying the wet documents and reassembling them. Clerselier then refused to show them to anyone, but something compelled him to let the young German see them, if only very briefly—he imposed a strict time limit on how long he could look at them. As it turned out, this young man was Gottfried Wilhelm Leibniz (1646–1716), a brilliant mathematician who then managed to decipher and copy Descartes's writings within the little time allotted him by Clerselier. Leibniz's copied notes now reside in an archive in Hanover, Germany.

GOTTFRIED LEIBNIZ

Gottfried Leibniz was born on July 1, 1646, in Leipzig, in a Germany that had been devastated by the Thirty Years' War. Until he was four years old, Swedish troops remained in a garrison, holding the city. His father, Friedrich Leibniz, taught moral philosophy at the University of Leipzig, and his mother, Catharina Schmuck, was the daughter of a law professor. In 1652 Leibniz's father died, and the six-year-old was sent to the Nicolai School, where he learned Latin—a normal practice among the European elite.

Young Leibniz retained access to the vast library left behind by his father, and it was here that he received his real education. He read voraciously the works of classical Greece and ancient Rome, and, according to the French historian Yvon Belaval, the budding child-scholar was able to use combinatorial reasoning to decipher the meaning of the words and sentences of the Latin language.[1] This deep understanding of language may have helped him years later to derive the key ideas of his masterpiece, *De Arte Combinatoria*. Guessing, defining, and combining the words of an unknown language as he deciphered their meaning had led Leibniz to combinatorial analysis—today a major component in modern probability theory.

Leibniz also developed a great appreciation for the life, customs, writings, and history of the ancients, which became evident throughout his adult life as he developed an interest in the historical continuity of ideas from antiquity to the present. By age twelve he was fluent in ancient Greek, having deciphered that language as well. He read Plato and Aristotle, which fueled his interest in logic and the foundations of pure reason. He also developed an interest in theology and its polemics. Thus we see in these adolescent investigations the roots of Leibniz's future groundbreaking work in theology, philosophy, and mathematics.

In 1661, at the age of fourteen, Leibniz enrolled at the University of Leipzig, where he studied the works of his contemporaries, including Thomas Hobbes, Francis Bacon, and Galileo. He also took courses on rhetoric, as well as Latin, Greek, and Hebrew.[2] In 1663 he presented his thesis, *De Principio Individui*, which dealt with ideas of individuation and totality, and that summer he took courses in jurisprudence, medieval history, and mathematics. He chose to concentrate his studies on law. The following year, Leibniz lost his mother, which was a devastating blow from which he took a long time to recover. Slowly, he returned to his studies and pursued the idea that proof in legal matters should rest on mathematical or logical evidence and its analysis—an approach that continues today with the use of scientific and statistical reasoning in criminal cases whenever such evidence and its analysis are possible.

In 1666, after receiving his master's degree in law, Leibniz published his treatise *De Arte Combinatoria*, an in-depth study of permutations that identified ideas as the basic elements of a logical system. Leibniz's point of view was that all concepts in the world are combinations of simple ideas—a notion that anticipated the *monads*, which would make Leibniz famous later on. Akin to atoms, monads are the eternal, indivisible building blocks of the metaphysical universe and make up everything complex in all creation. Leibniz felt that these primary elements should be few in number and as simple as possible so that everything in the universe could be reduced to the combination of these basic elements. Relations among ideas, Leibniz argued, could be deduced from uncovering the ways in which the simplest concepts are combined. For example, an interval of numbers is obtained through the combination of its elements—the numbers themselves—and 3-D objects can be seen as combinations of intervals. Similarly, all sentences can be seen as combinations of words, the words as combinations of sounds, and the sounds as combinations of the letters of the alphabet. Thus, at age twenty, Leibniz already possessed the basic concepts of his philosophy.

Next, Leibniz moved to Altdorf, Germany, and earned a doctorate at the local university. Then the intellectual life beckoned him to nearby Nuremberg, where he is rumored to have encountered the Rosicrucians. He served for a while as the secretary of the alchemical society's local chapter, although later he would call alchemy a "deception." Whatever the case may be, this connection to the Rosicrucians unites him with Descartes, who had also been rumored to have been involved with the society—if only at the suggestion of his adversaries. Through his supposed involvement with the Rosicrucians and with alchemy and astrology, Leibniz made the acquaintance of Baron Johann Christian von Boineburg, the former councilor to the elector of Mainz, one of several German princes who together were charged with electing the Holy Roman Emperor.[3] Boineburg was a celebrated European statesman, and the young Leibniz was attracted to the possibility of working with him.

Recognizing the young man's genius, Baron Boineburg invited Leibniz to Frankfurt, where he employed him as his personal librarian. Sometime later Boineburg took Leibniz with him to Mainz and introduced him to the princely court. Leibniz was a Protestant, but he was so attracted to the idea of working for a prince that he entertained the thought of converting to Catholicism, although he never did follow through with this notion. Through Boineburg and his royal contacts, Leibniz began his apprenticeship in the field of politics.

At one point Leibniz wrote a paper about his idea of basing jurisprudence on logical principles and, through Boineburg, presented it to the elector of Mainz. The elector was duly impressed by the brilliance of the young lawyer, and this move helped seal Leibniz's appointment as councilor to the chancellery of Mainz. In 1668 Leibniz wrote a treatise arguing for the existence of God and the immortality of the soul entitled *Nature's Testimony Against the Atheists*, which hinted at his future grand scheme for unifying the religions of Europe.

Over the next few years, Leibniz turned his attention away from politics, devoting himself to the study of mathematics, physics, and metaphysics. He wrote a number of papers and sent them abroad to the scientific academies in countries that had retained their supremacy in science and culture despite any religious wars they may have gone through—wars that had devastated his native Germany. In 1670 Leibniz sent a seminal paper on universal movement entitled *Theoria Motus Abstracti* to the French Academy of Sciences—one of the most important scientific bodies in Europe and, at the time, the arbiter of what science was and where it was heading. At the same time, he sent the Royal Society in London a paper entitled *Theoria Motus Concreti*, which applied to concrete, rather than abstract, notions of movement in mechanics. Leibniz also commenced a correspondence with various intellectuals in France and Britain, hoping one day to turn his new ties into a permanent, paying position in either of these two countries.

Leibniz read Descartes and quickly made up his mind that the French philosopher's ideas were anathema to him. "I am nothing less than a Cartesian," he wrote to the French philosopher Antoine Arnauld, a man who had written diatribes against Descartes.[4] What Leibniz objected to in Descartes's philosophy was the latter's separation of body and soul. In his own writings, which were then attacked by the Cartesians, Leibniz sought to find the soul within the physicality of the body. These debates enhanced his reputation as a philosopher, and he extended his correspondence with European intellectuals, including the famous Dutch philosopher Spinoza.

Leibniz was different from many other geniuses throughout history in that he achieved so much in so many

The theologian Antoine Arnauld (1612–94) carried on a lively correspondence with Leibniz regarding the philosophy of René Descartes.

disparate fields. While battling the Cartesians on philosophy, he also invented a machine that could perform addition, subtraction, multiplication, division, and calculate square and cubic roots. This device of immense genius and value was seen as superior to a similar mechanical calculator created by Blaise Pascal. Leibniz also invented lenses, air pumps, and a nautical navigation instrument—all while serving the princely court of Mainz and Baron von Boineburg. It has been said that Leibniz lived several lives; each of his occupations gave him the equivalent of a full lifetime of knowledge. Some lamented this fact, believing that he wasted his genius for mathematics by not concentrating more on that field.[5]

Then Leibniz developed the strange idea of trying to manipulate the king of France. Sometimes even great minds can hatch senseless schemes. This was one such example.

A HAREBRAINED CONSPIRACY

Leibniz was a product of the Thirty Years' War and, hence, sensitive to the existence of religious differences among the peoples of Europe. He viewed these differences as eternal causes of conflict, so he came to the idea that Protestantism and Catholicism should be united in some way. In this respect, he was especially worried about the very powerful king of France, a Catholic whom Leibniz saw as a threat to European peace because of his supposed desire to attack Protestants, which were numerous in German principalities as well as in Holland.

Four decades earlier, in 1628, Louis XIII killed off many of his own Protestant citizens in the siege of La Rochelle, witnessed by Descartes. Understandably, Leibniz feared that Louis XIII's successor, Louis XIV—a.k.a. the Sun King—posed a threat to Protestants outside his realm. Leibniz conceived a political scheme to divert the French king from what he thought was a design on Protestant nations and regions. Would it be possible, Leibniz asked himself, to tempt Louis XIV to launch an attack on Muslim Egypt as a way of countering the increasing influence of the Ottomans? If France were to use its might against Egypt, he reasoned, it would be unlikely to launch attacks on European Protestants.

Boineburg had been aware of Leibniz's idea that the king of France might be induced, if approached correctly, to change direction and turn his military might against the Muslim infidels of Egypt instead of against his Christian brothers in northern Europe. In 1672 the baron sent his librarian and budding diplomat to Paris with the charge of finding a way to approach Louis XIV and propose his plan. Leibniz's other mission was to arrange for the education of Boineburg's son in the French capital. Thus, with Boineburg's son under his charge, Leibniz happily left for Paris, a city he had always wanted to see and hoped to live in.

Hyacinthe Rigaud's 1701 portrait of Louis XIV, whom Leibniz tried to coax into invading Egypt, appears against the background of a battle scene painted by Charles Parrocel.

The French were enemies of the Germans—they occupied German cities, as did their allies, the Swedes—so the audacity of a German statesman sending a minor diplomat to France to convince the king to invade Egypt strikes us today as naive and childish. In fact, Leibniz's complete plan was even more outlandish. According to his strategy, the French, after conquering Egypt, could also be induced to invade the rest of North Africa and the Levant, a region that included several nations on the east coast of the Mediterranean. Additionally, he dreamed, Sweden and Poland could invade and "civilize" Siberia, the Crimea, the Black Sea, and the Sea of Azov; England and Denmark could invade all of North America; Holland could take over the East Indies; and Spain could consolidate its control over all of South America, without Portugal—a nation Leibniz considered unimportant. Thus, according to Leibniz, the entire world could eventually be occupied by European nations, whose religions would be unified under the rubric of "Christianity," without distinction between its denominations.

The silly plan came to nothing before it even had a chance of being presented. Leibniz had barely arrived in Paris when, on May 6, 1672, Louis XIV declared war on Protestant Holland. Leibniz's master, the Prince of Mainz, offered to serve as mediator between France and Holland, but Louis XIV rebuffed him, and French troops invaded Holland.

Leibniz noted that Louis XIV was universally hated in Europe, and he saw the king as an aggressive leader consumed with the desire to attack his neighbors, while "a large part of his people eats only once a day."[6] Although he had never met the king, the young German diplomat was able to make connections with many French statesmen, scientists, and intellectuals. Among them were the Duke of Chevreuse and the philosopher Antoine Arnauld, with whom he already had an established correspondence. He also met the Prince of Condé, who expressed great interest in Leibniz's idea of a religious unification of the continent.

Not only did Louis XIV invade Holland—obviating Leibniz's mission in Paris—but in December 1672 Leibniz's patron Boineburg died, and a few months later, in early 1673, so did Prince Johann Philip, the elector of Mainz. Leibniz continued to tutor Boineburg's son until sometime in 1674, but he refused to return to Germany, where he had no employment prospects left.

After failing to get an audience with Louis XIV, Leibniz settled in Paris for four years. Now free to do whatever he wanted, he chose to focus on mathematics and the physical description of the universe, expanding upon the physics of Galileo and his contemporaries. Leibniz also drew upon the ideas of Eudoxus, Archimedes, Descartes, and Fermat, while developing the key mathematical theory we know as calculus.

In 1673 Leibniz took a trip to Britain, remaining there for three months. In England he met the physicist Robert Boyle, as well as the mathematicians Christopher Wren, John Pell (of the famous Pell equation), and the German intellectual Henry Oldenburg, who served as secretary of the Royal Society in London. That same year, Leibniz published the philosophical treatise *Confessio Philosophi* (A Philosopher's Creed).

Leibniz returned to Paris invigorated by the mathematical and physical ideas he had discussed with British mathematicians, but he soon realized that without employment his days in the lovely French capital were numbered. He needed a new patron, and he soon found one in the person of Duke Hans Friedrich of Hanover.

Leibniz is depicted in this ca. 1700 portrait by Johann Friedrich Wentzel.

The duke agreed to bankroll Leibniz's stay in Paris. Leibniz was now free to reside in the city and turn his full attention to mathematics. Here, under the care and guidance of Christiaan Huygens, a brilliant mathematician, astronomer, and physicist seventeen years his senior, Leibniz was to become a first-rate mathematician. As Leibniz wrote in his memoir:

> It was thus that Huygens who, I think, saw in me more than there was, brought me with gentility a recently published copy of his book on the pendulum [*Horologium Oscillarium,* 1673]. This was for me the beginning of the opportunity for a deeper study of mathematics. While doing so, Huygens realized that I had not an understanding of the concept of center of mass and he gave me a brief description, noting that [Blaise Pascal] had treated it remarkably well.[7]

Leibniz went back to the ideas of his younger days and tried to apply his combinatorial analysis to a wider range of problems in mathematics. In 1674 he was studying a figure in a work of Pascal when he had a brilliant insight into functions and rates of change. A year later, this insight led him to propose a coherent and complete system of computing derivatives, which are instantaneous rates of change of functions; and a general way of finding areas under the curves of functions—i.e., integrals. This system comprised what later became known as infinitesimal calculus.[8] Leibniz even introduced the universal notation we use in calculus today: the differential symbols *dx* and *dy*, as well as the integral sign, ∫, an elongated *S*, from the Latin word for sum: *summa*. The integral operation is, in fact, a continuous form of summation. Leibniz introduced this sign for the first time in a paper he wrote in November of 1675.

THE CALCULUS CONTROVERSY

Once Leibniz published his calculus, he was fiercely attacked by British mathematicians, who accused him of stealing from the works of Isaac Newton. Some of them knew that Newton had been working on the same calculus methods for years; others even claimed that all Leibniz had done was to "steal the ideas of Descartes." Leibniz became convinced that the only way to defend himself was to read everything Descartes had written. It was at this point in time that he found Claude Clerselier, the keeper of all Descartes's works. It was also at this time that, by reading Descartes anew, he became a convert to the Cartesian approach to understanding the universe.

Reading Descartes, Leibniz underwent a strange transformation. While the unpublished writings of Descartes brought Leibniz into the fold of Cartesianism, the mathematics of Descartes also held a special attraction for him. Leibniz saw that Descartes had developed mathematical ideas and methods similar to those that he himself had been entertaining. In particular, Leibniz recognized that Descartes was so gifted as a mathematician that he could "use" calculus without being in possession of a complete methodology. In other words, Descartes had been able to find the *derivative* of a mathematical function without, perhaps, recognizing that he was doing completely new mathematics.

Leibniz's perusal of Descartes's manuscripts convinced him that no one could reasonably claim that he had stolen the calculus idea from Descartes. But a question remained in the minds of many people: Did Leibniz meet John Collins, a friend of Newton's, during his first visit to Britain in 1673 and, through Collins, learn about Newton's work on calculus? Or did this meeting take place in 1676, after Leibniz's own published formulation of calculus? As one historical testimony maintains, "It seems that he only met Collins on his second voyage."[9] But the phrase "it seems" reveals historical doubt. The British were quick to attack Leibniz as a plagiarist, and the Germans defended him. Before long, the controversy about who had invented calculus took the nationalistic tones of England versus Germany.

French author Bernard le Bovier de Fontenelle, who wrote Leibniz's eulogy, concluded, "If it was a theft, then it was a theft that only Leibniz could carry out."[10] In fact, the calculus of Leibniz is more comprehensive and fertile than Newton's calculus. Whereas Newton approached the idea of the variation in a function in terms of bodies in motion and the concepts of speed and acceleration, Leibniz used the idea of mathematical infinitesimals in his approach. In 1684 Leibniz published a work entitled *Nova Methodus Pro Maximis et Minimis* (*A New Method for Finding Maxima and Minima*), the main application of differential calculus. Newton's work on calculus, which he termed "the method of fluxions and fluents," appeared in print only three years later, in 1687.

Leibniz strove to apply his new mathematics to metaphysics and theology. The infinitesimals he invented for his work—or, rather, adapted from the works of the ancient Greeks—held mystical powers in his eyes, and he hoped to use them in metaphysical investigations. While Newton, too, was a religious man, his calculus was purely a response to the needs of physics rather than anything metaphysical. Ultimately, both Leibniz and Newton are equally credited with independently developing the modern theory of calculus based on the work of Eudoxus, Archimedes, Fermat, Descartes, and other mathematicians.

But this was not the understanding in the late seventeenth century, as the virulent controversy raged in Europe. When Newton, whom we will soon meet, heard that Leibniz had published a paper on calculus and had previously made contact with British mathematicians familiar with his own work, he became testy and deeply suspicious. He immediately wrote Leibniz a letter about his own work and dispatched it to him through Henry Oldenburg at the Royal Society. Then he anxiously waited for Leibniz's answer and explanation as to how, exactly, he had derived his method and whether he had any knowledge of Newton's own work. But as fate would have it, the letter took a long time to arrive—about six weeks. Thus, when Newton received Leibniz's response, he assumed that Leibniz had taken a very long time to craft an answer and became even more suspicious.

In Hanover, Leibniz received a second letter from Newton. Because the letter was forwarded to him from his previous address, there was once again a long delay in his response. Newton's letter was written on October 24, 1676, while Leibniz was traveling, and arrived in Hanover in June 1677—eight months later! Leibniz replied immediately, explaining that he had developed his methods all on his own with no input about Newton's work, but this second delay further convinced Newton that Leibniz was stalling. The controversy raged on, fanned by Newton's anger.

In 1711 the Scottish mathematician John Keill published a paper in the *Transactions of the Royal Society of London* that squarely accused Leibniz of plagiarizing Newton's work. Upon reading it, Leibniz wrote the Royal Society saying that he had never heard about Newton's version of calculus before the publication of his own calculus. Keill wrote him back saying that Newton's letters gave ample proof of plagiarism. Leibniz then wrote the society, asking that his name be cleared. In response, a committee of inquiry was established. As we will soon see, the committee's report was written by none other than Newton himself.

In the meantime, Leibniz had to give up his beloved Paris and return to Germany. By late 1676 he was out of money, and Huygens's attempts to help him gain acceptance into the French Academy of Sciences were in vain. Ironically, Leibniz would become the force that led to the founding of other academies of science in Europe, including the Prussian and Russian academies. Although the prime minister of Denmark offered Leibniz a position as councilor to the Royal Danish Court, Leibniz was not interested in relocating to Denmark. Before returning to his homeland, he spent a week in London, where he met with Boyle, Oldenburg, and Collins. Then he spent a month in Holland, where he met with Spinoza. In December of that year, Leibniz finally returned to Germany to assume the post of librarian, diplomat, tutor, court philosopher, and counselor to the Duke of Hanover.

His employer, Johann Friedrich, was chiefly concerned with reorganizing his army. The duke had recently converted to Catholicism and now zealously oversaw the adoption of his new religion by all his subjects. Leibniz admired his sovereign for his faith but disagreed with his politics. In fact, the Duke of Hanover was the military ally of Louis XIV, which made Leibniz uncomfortable. Still, ever since France's attack on Holland a few years earlier, the House of Hanover had remained nominally neutral.

Leibniz wrote a thesis about the sovereignty of the German principalities and argued for a German federalism. He continued to push for a union of the religions of Europe and for a federalism of the entire continent under the supreme leadership of the Holy Roman Emperor—controversial ideas that nevertheless contributed to his appointment to the chancellery of Hanover in 1678. Later on he became involved in manufacturing and heavy industry. He tried to negotiate for the export of metals mined in Germany's Harz Mountains and proposed to the Austrians that streetlamps in Vienna be fueled by rapeseed oil, which would save money and provide efficient lighting.

Leibniz remained a bachelor all through his life. When he was fifty, he finally proposed to a woman, but she took so long to consider his proposal that he decided to withdraw it. He died in Hanover at age seventy. Unlike Newton, who was buried in a place of honor—Westminster Abbey—Leibniz was put to rest in an unmarked grave. Only his secretary and a few bystanders attended the funeral.

ISAAC NEWTON

Like his mortal rival Leibniz, Newton, too, remained a bachelor throughout his life. And like Leibniz, he was also a man of many great achievements, each of which was monumental.

Isaac Newton was born on Christmas Day in 1642, the year that Galileo died. He came from a farming family in Woolsthorpe, in the county

Isaac Newton was born in this house, Woolsthorpe Manor, in Lincolnshire, England. Today, it is maintained by the National Trust and is open to the public.

of Lincoln, England. Newton was a premature baby, and his mother once described him as having been so small at birth that he could fit inside a quart mug. Two neighbors who visited his mother shortly after his birth and then went away for a short while said they had expected him to be dead when they returned; but the baby survived.[11]

Newton's father, also named Isaac, died at age thirty-seven and never had the chance to get to know his new son. When Isaac was three years old, his mother, Hannah Ayscough, married a much older man named Barnabas Smith, the minister of the church in a neighboring village. Upon remarrying, she left baby Isaac to be raised by his grandmother, Margery Ayscough. Hannah then moved away and, with her new husband, had three children.

As a child living with his grandmother, Newton invented and built many mechanical toys, including a small flour mill. In one prank he put lanterns on kites and flew them up in the night sky to scare the villagers. Newton was on his way to becoming a farm boy. Later, when he was ten, his mother returned to the village after Barnabas Smith had died, and Newton was expected to stay in the village to assist her with chores.

In 1659 Newton's mother yanked him out of the free grammar school in Grantham, which he had been attending. He then lived at home, part of a large household that included his mother, his grandmother, and his half siblings. But Newton was rescued from the unhappy fate of a farm boy by his uncle, William Ayscough, who had graduated from Cambridge and who recognized the boy's promise. He convinced his sister, Newton's mother, to send her bright son to be educated at that great university. In order to prepare for Cambridge, Isaac was allowed to return to the Grantham School to finish his education.

While attending school, Newton lodged with the family of the village apothecary, a Mr. Clarke, and fell in love with Clarke's stepdaughter. The two got engaged, but in June 1661 Newton left for Cambridge and was soon so immersed in his studies and reading that he never married her— or anyone else. Newton supported himself as a student by doing menial work at the university but, nevertheless, excelled in his studies. His mathematics professor, Isaac Barrow—the first Lucasian Professor at Cambridge—taught geometry and used his own methods for finding tangents to curves and areas of geometrical figures. Not only would Newton surpass him by inventing calculus, a methodology for performing these very tasks in a systematic way, but he would eventually replace Barrow as the Lucasian Professor.

Newton was equally occupied with theology and alchemy, the forerunner of our modern science of chemistry. He was a religious man and spent much time during those years trying to make sense of the

Newton spent nearly forty years at Trinity College, Cambridge, first as a student and then as a professor. The rooms in which he lodged are located on the first floor in the low center building just to the right of the entrance gate.

prophesies of Daniel and the Apocalypse. He also tried to determine the date of the creation of the world based on biblical writings, which he interpreted literally. At the university, Newton studied hard, but he also found time to relax, going to taverns and playing cards.[12] Because Newton was secretive about his work, we don't know about any discoveries he might have made during his period as an undergraduate. In 1664 he earned his bachelor's degree, and then began his most fecund period.

DESCRIBING WHAT HAPPENED NEXT, Newton famously said, "If I have seen a little farther than others it is because I have stood on the shoulders of giants." Presumably, the giants on whose work the modest Newton had relied included Descartes, Kepler, and Galileo. Cartesian logic led him forward, as did Descartes's mathematical work. Galileo's investigations of falling bodies, the pendulum, and other physical phenomena inspired his own theory of gravity. And Kepler's laws of planetary motion were later abstracted by Newton, making them corollaries of his laws of universal gravitation. Standing on the shoulders of his giants, Newton saw much farther than anyone. He was not a man of universal interests, as was his contemporary and calculus cofounder Leibniz, but in the realm of physics and mathematics, Newton's intellect was supreme.

In 1664 Britain was ravaged by bubonic plague, and Cambridge University was closed down. Newton left for Woolsthorpe, where he spent two years alone, thinking about the universe and its laws. It was here, for the purpose of explaining the laws of gravitation he deduced from physical investigations, that Newton invented calculus. Newton viewed variables as flowing quantities, and to describe this flow—the rate of change of a quantity with time—he devised what he called the "method of fluxions." Before he achieved this great breakthrough, he generalized the binomial theorem—the rule for operations such as squaring the sum of two quantities—by extending it to higher exponents and also to cases where the exponent is not a positive integer. In such cases the series is infinite, and proving the result is much harder.

Newton's law of universal gravitation states that any two particles of matter attract one another gravitationally with a force that is proportional to the product of their two masses and inversely proportional to the square of the distance between them. The constant of proportionality in the equation is known as Newton's constant, *G*. Newton also developed laws of motion, which include the following:

1. Every body will continue in its state of rest or inertia (unaccelerated motion) in a straight line unless acted upon by a force.

2. The rate of change of momentum (mass times velocity, in Newtonian physics) is directly proportional to the force acting on a body and inversely proportional to the mass of the body.

3. For every action, there is an equal and opposite reaction.

In his laws of motion, Newton neatly outlined the relationship between the acceleration, velocity, and position of an object (i.e., acceleration is the rate of change of velocity, and velocity is the rate of change of position). These rates of change became the derivatives of differential calculus, and in order to measure the distance traveled by an object whose velocity changes over time, Newton invented integral calculus. Further, understanding that these two operations—finding a rate of change (i.e., computing a derivative), and finding a distance or area or volume (i.e., computing an integral)—are opposite operations, he formulated the fundamental theorem of calculus.

In 1684, in order to settle a dispute with fellow scientist Christopher Wren, Edmond Halley of the Royal Society—of which Newton was a member—asked him which law of attraction would explain the elliptical orbits of the planets. Newton replied that it would be an inverse-square law—that is, a law by which the force of attraction is decreased according to the square of the distance between a planet and the sun. "How do you know that?" asked Halley, to which Newton answered, "Why, I have calculated it."[13] In fact, Newton had performed a calculation in 1666 demonstrating that his law of universal gravitation led directly to Kepler's laws of planetary motion. But he did not publish it until eighteen years later.

Why did he wait so long between derivation and publication? Although Newton understood that every point in a solid body exerted the same gravitational force on every point in the second body, it was not clear to Newton how to perform the computation for a large number of points. Eventually, he figured out that if he assumed that a body's mass was concentrated at its center, he could calculate the gravitational force on another body.

In 1667 Newton was elected fellow of Trinity College, and two years later he was made Lucasian Professor of Mathematics at Cambridge, succeeding Isaac Barrow. Newton gave lectures on optics that included his own discoveries in the field. In contrast to the theory of Christiaan Huygens and Robert Hooke, which held that light was a wave, he espoused a corpuscular theory of light. Today we know that light is both a particle and a wave, as quantum theory has taught us.

In 1668 Newton built a new kind of telescope that used a reflecting mirror. (Such telescopes are still called Newtonian.) He then used his reflecting telescope to look at the night sky and discern the Galilean satellites. That year, Dutch mathematician Nicholas Mercator had produced a calculation of an area under the hyperbola using an infinite series that was closely related to Newton's calculus. Since Newton's method had not yet been published, the work of Mercator provided Newton with the impetus to circulate his own work among mathematicians at Cambridge.

In 1672 Newton was elected to the Royal Society. There, he communicated to the world of science his work on telescopes. He also read papers in front of the Society about his "particle" theory of light, prompting an argument with Robert Hooke about the nature of light. Other members sided with Hooke's wave theory, and Newton's letters to the Royal Society about the issue became angrier and angrier as time went by. In a letter dated November 18, 1676, he wrote, "I see a man must either resolve to put out nothing new, or become a slave to defend it."[14]

Newton's friend Edmond Halley eventually persuaded him to put his ideas to paper, lest credit for them go to other scientists and mathematicians. Thus in 1684 Newton began writing

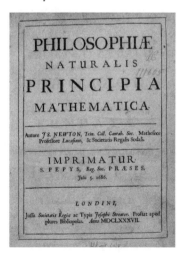

The title page of the first edition of Newton's *Principia Mathematica* shows that it was published in London in 1687.

In the dispute about who first discovered calculus, Newton prevailed over Leibniz—at least during his lifetime. Today, however, we recognize the contributions of both men to its development. This engraving depicts Newton at his prime.

his masterpiece, *Philosophiae Naturalis Principia Mathematica* (*Mathematical Principles of Natural Philosophy*). Writing the *Principia* was a major undertaking, and it took a toll on its author, who worked for hours on end, sleeping little, waking up to write while still in bed, and eating poorly. It was a vast amount of knowledge, acquired over decades, that now had to be put down on paper.

Newton wrote about astronomy, physics, and the mathematics he had invented. But the main focus of the book was the dynamic structure of the universe—i.e., the solar system, with planets revolving around the sun, according to Newton's law of universal gravitation. Newton's calculus methods underlie the dynamics, but he reduced the calculations to geometrical arguments that could be easily understood. Newton deduced Kepler's laws of planetary motion from his own law of universal gravitation, and he showed how to calculate the sun's mass.

Newton also managed to show that comets' return to Earth's vicinity could be predicted by his gravity law, and he explained that Earth's flatness at the poles was due to its rotation about its axis. Other observations tied to what he called the System of the World, under the rubric of his all-powerful law of universal gravitation, include tides as a result of the moon's gravitational pull, the precession of the equinoxes, and the variation in the weight of an object with latitude. Newton's *Principia* became part of the university curriculum in England around the turn of the eighteenth century, being taught at both Cambridge and Oxford. Meanwhile, news of his groundbreaking scientific and mathematical discoveries spread throughout the world.

FROM 1689 TO 1690, Newton was a member of the Convention Parliament in London, representing Cambridge University. In 1696 he was elected Warden of the Royal Mint in London and, three years later, promoted to the very prestigious post of Master of the Royal Mint, in recognition of his service to society in the creation and dissemination of knowledge. Income from his position as the head of the mint enhanced his wealth, and while many others might have rested on their laurels, Newton worked hard at the mint, improving methods to combat counterfeiting and making the minting process more efficient as new coins were brought into circulation.

While working in London, Newton suffered from a severe inability to sleep, and he developed poor eating habits. News of his illness spread throughout Europe, bringing wishes for a quick recovery from many. On hearing the news that Newton was well again in 1693, Leibniz wrote to a friend saying how relieved he was. But that same year, Newton heard that calculus was spreading through the continent of Europe and that it was credited to the work of Leibniz rather than himself. Thus began what historian Eric Temple Bell calls "the most shameful squabble about priority in the history of mathematics."

On June 20, 1688, Newton wrote to Halley that science was "such a litigious Lady, that a man had as good be engaged to lawsuits, as to have to do with her."[15] He was becoming consumed with his fight against Leibniz and was determined that he, not the German mathematician, be given full credit for inventing calculus.

In 1703 Newton was elected to preside over the Royal Society, and thereafter he was reelected to this position of power and prestige every year for the rest of his life. In 1705 he was knighted by Queen Anne. Newton used his new powers against Leibniz: he appointed the committee that investigated the priority over calculus, and (anonymously) authored its final report, finding in his own favor.

But the dispute over calculus was never settled, and today we recognize both Newton and Leibniz jointly as the co-inventors of calculus. Leibniz's dx and dy notation eventually won out over Newton's $\dot{x}$ and $\dot{y}$ "fluxions," however. (But Newton's "dot" notation is still used in physics and other applications of calculus. In truth, Newton's calculus was both cumbersome and less rigorous, while Leibniz's notation and ease of use made his form of calculus useful.) Newton's calculus, though, was developed before that of Leibniz, albeit not published early enough. Regardless of who is rightfully credited with its discovery, European mathematicians found in calculus fertile ground for developing and extending much beautiful and useful mathematics in the years to follow.

GENIUSES OF
THE ENLIGHTENMENT

The seventeenth century brought immense advances in mathematics, which took the pioneering work of the ancient Greeks and catapulted it into the modern age, culminating in the birth of the calculus. In the century that followed, calculus was taken to higher levels of understanding, application, and abstraction, and mathematical analysis as we know it today was born. Imaginary numbers came of age, giving rise to complex analyses, and new geometries that went far beyond Euclid's imagination unseated the ancient scholar and his predecessors.

Sixteenth-century mathematics flourished in Italy, and mathematics of the seventeenth century found fertile ground

Several key mathematical geniuses of the Enlightenment, including members of the Bernoulli family and Leonhard Euler, studied at the University of Basel, seen in this modern photograph.

in France and Britain. By contrast, leading mathematicians of the eighteenth century were mostly Swiss and German, though, having migrated from one part of the continent to another, they may properly be considered pan-European.

THE BERNOULLI DYNASTY

In Basel, Switzerland, there lived a family of gifted mathematicians called the Bernoullis. The Bernoulli family had originated in Spanish-controlled Antwerp, but in 1576 they fled religious persecution. Seven years later they moved to Basel. This most unusual family begat a dozen great mathematicians.[1] Two of them—brothers who lived in the late seventeenth and early eighteenth centuries—and their descendants play a role in our story.

Nicolas Bernoulli, the father of these two brothers, was a spice merchant and prominent citizen who sat on the Basel city council; his wife came from a wealthy family of Swiss bankers. He was not interested in his sons pursuing mathematical careers. He wanted Jacob (also known as James or Jacques), the older brother, to be a cleric and the younger, John (also known as Jean or Johann), to be a merchant or a doctor. The father piled obstacles in the way of the two brothers pursuing mathematics, but to no avail.

Jacob was forced by his parents to study philosophy and theology at the University of Basel in preparation for a career as a minister. He received his master's degree from the university in 1671 and five years later was awarded a license to practice theology in Switzerland. Throughout his education, however, he read mathematics—his true passion in life—on the side. Jacob then moved to Geneva, in the French part of Switzerland, and from there to France, where he concentrated his studies on the mathematics and philosophy of Descartes. On a visit to Britain, he met Newton's associates Boyle and Hooke and thereafter maintained an extensive correspondence with them and many other mathematicians and physicists he met during his travels. Throughout this period, and to his parents' dismay, Jacob supported himself by working as a mathematics tutor.

After studying all the works of Leibniz from 1684 to 1686, Jacob Bernoulli became convinced of the great power of the new calculus. In fact, it is he who suggested to Leibniz that the term *integral* denote the area under the curve represented by a function. Jacob extended the range and depth of calculus in a number of papers he published in professional journals. Like Newton, he investigated infinite series and was able to prove that the harmonic series $(1/n)$ is divergent—although, according to him, his bother John first made the discovery.[2] He also showed that the series of the reciprocals of the Pythagorean square numbers—i.e., $1/n^2$—is convergent, although he did not determine what it converged to.

Jacob also proved the interesting Bernoulli Inequality: $(1 + x)^n > 1 + nx$, where x is a real number greater than -1 and not 0. The relationship also works for non-integers. In 1690 he published the first paper to reference the integral, substituting Leibniz's *calculus summatorius* with *calculus integralis*. He also proved the law of large numbers used in probability theory.

When John Bernoulli read Leibniz, he was so taken with his calculus that he spent two years writing about it. Not having a profession, he went to Paris—against his father's better judgment—to work as a tutor for a young French nobleman, the Marquis de l'Hôpital (1661–1704). John taught the marquis about Leibniz's calculus, but because he needed money after he left Paris, he agreed to keep sending the marquis his mathematical developments—to be disposed of as the marquis wished—for a continuing salary. Thus in 1694 Bernoulli sent l'Hôpital

This Swiss stamp features a likeness of Jacob Bernoulli painted by his brother Nicolas, along with Bernoulli's law of large numbers.

a rule that states that if two functions possess derivatives and are equal to 0 at the same point, and if the limit of the ratio of the two derivatives exists at that point, then the limit of the ratio of the two derivatives at the point is equal to the limit of the ratio of the two functions at that point. This is a standard rule in advanced calculus, and all students learn it under the name l'Hôpital's rule. The Marquis de l'Hôpital used his agreement with Bernoulli to publish the rule in his book *Analyse des Infiniment Petits pour l'Intelligence des Lignes Courbes* (Analysis of the Infinitely Small to Understand Curved Lines) without giving its discoverer any credit.

John was a fervent supporter of Leibniz and aggressively pushed for Leibniz's priority in his dispute with Newton. Though his acquiescence in sending his works to l'Hôpital may suggest otherwise, he was actually a very competitive and combative person. In fact, when his very gifted son Daniel Bernoulli (1700–82) won a prize for mathematics given by the French Academy of Sciences, John drove him out of the house because he had applied for the same prize and lost.

John had two other sons, named Nicolas and John. All three were superb mathematicians, but other members of this extended family achieved renown in mathematics as well.

THE PETERSBURG PARADOX

Throughout his life, Leibniz lobbied for the founding of royal academies. In France the Royal Academy of Sciences—today known simply as the Academy of Sciences, having survived the fall of the monarchy during the French Revolution—is the official body to which top French (and a few foreign) scholars belong. Germany and Russia, too, had very active royal academies in those days. The monarch— Catherine I in the case of Russia, and Frederick the Great in the case of Prussia—each established a royal academy to attract the top thinkers to the employ of the crown. Not only were royal academies the most prestigious research institutions, but they were also the leading places in Europe for doing research in science and mathematics in particular. Universities were more interested in teaching than in research, and many sciences were too new to be supported strongly there. Thus for top scientists and mathematicians the royal academy was the best place to conduct research in one's field.

In 1725 the Bernoulli brothers Daniel and Nicolas (John's sons) were invited to the Royal Academy of Saint Petersburg. They were both interested in the theory of probability, and together they worked on the idea of mathematical expectation, as illustrated by the following question: If an investment has a 50 percent chance of earning $1,000 and a 50 percent chance of losing $200, how much is it worth? According to the laws of the theory of probability, the answer can be determined by multiplying the values by their probabilities and adding the results:

$$0.5(1,000) + 0.5(-200) = \$400$$

This calculation of the expected value presents a reasonable long-term outcome for a series of trials. In other words, if continuously faced with a long sequence of repetitions of this investment opportunity, one will earn an average of $400 per investment. Expanding on this theory, Daniel and Nicolas Bernoulli came up with a scenario that led to the famous Petersburg Paradox.

The example Daniel and Nicolas developed involves Peter and Paul. Peter tosses a coin continuously until a head appears, and the number of tosses correlates to the amount he owes Paul in the following way. If a head appears on the first toss, Peter pays Paul two crowns. If the result is a tail, Peter throws again, and if a head appears, Peter pays Paul four crowns. If two tosses are tails and the third one is a head, Peter pays Paul eight crowns; and so on. The question is: For Paul, what is this game worth? First of all, we need to represent the probability of tossing a head after x number of tosses with a function. The initial probability of tossing a head is $\frac{1}{2}$, but the probability of tossing a head after tossing a tail is only $\frac{1}{4}$. The probability of tossing a head on the third toss is $\frac{1}{8}$, and so on, so we can represent the probability of tossing a head after x tosses with the function $y = (\frac{1}{2})^x$. The amount of money Peter owes Paul after each toss can be represented by the function $y = 2^x$. Further, if we represent the expected value as a summation of the factors of these two functions multiplied together, as the rule for expected values requires us to do, we find that the game has infinite worth to Paul:

$$(\tfrac{1}{2})2 + (\tfrac{1}{4})4 + (\tfrac{1}{8})8 + (\tfrac{1}{16})16 + \ldots = 1 + 1 + 1 + 1 + \ldots = \text{infinity}$$

This expectation is an apparent paradox because the game won't last to infinity. In fact, simulations have shown that the amount earned by Paul is usually less than five crowns.

The Petersburg Paradox occupied European mathematicians for many years and, thanks to the Bernoullis, led to the distinction between actual expectation and "moral expectation"–i.e., the fact that the actual "expected value" is not always mathematical; rather, it may be affected by people's attitudes to money and risk–that is, whether they are risk-averse or risk-seeking.

LEONHARD EULER

Daniel and Nicolas Bernoulli were friends of another extremely gifted Swiss mathematician from Basel who was employed at the Saint Petersburg Academy: Leonhard Euler (1707–83). Euler (pronounced "oiler") has been called "the most prolific mathematician in history." His contemporaries called him "analysis incarnate," and in the words of the French scholar Francois Arago, "Euler calculated without apparent effort, as people breathe or as eagles sustain themselves in the wind."[3] Euler's works comprise 73 volumes of collected papers and 886 books and articles.

Leonhard Euler was the first of six children born to the family of a Calvinist minister in Basel, Switzerland. At age seventeen he earned his master's degree from the University of Basel, where he studied mathematics, physics, astronomy, languages, and medicine. His father, Paul Euler, a minister, was also an amateur mathematician. In fact, he had studied mathematics under Jacob Bernoulli. At a young age Leonhard learned mathematics from his father, but because his father was eager to have his son succeed him as pastor of the village of Riechen, young Euler also studied theology and Hebrew. At the University of Basel, he excelled so much in mathematics that he caught the attention of John Bernoulli, who began giving Euler a private lesson once a week.[4] John's children, Nicolas and Daniel, commenced a friendship with the young genius that would last a lifetime. When Euler's father insisted that he be groomed to take over as pastor of Riechen after graduating in 1724, the two Bernoulli brothers intervened, promising Euler's father that his son would be a far greater mathematician than clergyman. Though saved from a life of religious service, Euler would remain a man of faith, leading his large family in prayer and giving sermons to his children.

In 1727 the French Academy of Sciences offered a prize to the person who could design the best mast for a ship. It is believed that, other than small vessels sailing on Swiss lakes, Euler had never seen a ship—yet his design was the runner-up for the big prize and deemed the best from a mathematical-theory point of view. In later years he would win this prestigious award twelve times. In fact, throughout his life he saw the physical universe as nothing but an excuse to pursue mathematics—he was never actually interested in the applications that often instigated mathematical pursuits.

Euler applied for a professorship at the University of Basel but didn't get it. By then, his two friends were in Russia, working for Catherine I as researchers in her royal academy. The two wrote Euler constantly, informing him of their continuing efforts to secure him a similar appointment. In the meantime, Euler remained a student, taking more courses at Basel. At one point the brothers wrote their friend that an opening had come up in medicine. Euler then threw himself into the study of medicine and consequently was invited to take a position in medicine at the Saint Petersburg Academy in 1727. Euler rushed to Saint Petersburg, and when he arrived, there was so much confusion at the academy because of political issues, hires, and dismissals that no one noticed when the person invited to assume a medical position slipped unnoticed into the mathematics section. There, Euler thrived, and in 1733, after Daniel Bernoulli returned to his native Switzerland, he became the head of the department.

At age twenty-six Euler decided that he would make a home in Russia. He met a Swiss woman named Katharina Gsell, the daughter of the artist Georg Gsell, who had been recruited by Peter the Great of Russia to serve as curator of the Imperial Art Gallery in Saint Petersburg. The tsar had met Gsell in Amsterdam and, hoping to improve art in Russia, asked the

Swiss artist to help him acquire works of Dutch Masters. In 1734 Gsell's daughter Katharina married Euler. The couple had thirteen children, but only five survived to adulthood. Euler was so attached to his children that he would often work on mathematics while holding a baby and watching the other children play around him.[5]

Just over a year into his marriage, Euler went almost blind in his right eye. Though the problem may have been exacerbated by the extremely concentrated work Euler did over a period of several weeks, his vision problems did not slow the progress of his extensive research in mathematics. In fact, even when he lost most of the sight in his other eye many years later, in 1766, he still found ways to continue his prodigious work, relying heavily on his perfect photographic memory and imagination. At the height of his career, he would produce a mathematical paper within a few days, although legend has it that he would finish an entire paper between two calls for dinner.[6]

In light of the political turmoil in Russia, which caused difficulties for foreign residents, Euler decided in 1741 to accept an invitation for a position at another great royal academy of the time, that of Frederick the Great in Berlin. During his twenty-five-year stay in Germany, Euler undertook a massive amount of work and wrote nearly four hundred articles, but he did not get along well with the Prussian emperor. Apparently, the emperor wanted Euler to help him design a water-transport system for his summer palace of Sanssouci, outside the city, and Euler—a pure mathematician, not an engineer—failed at this task. A pious Swiss country boy who excelled in mathematics but could not

The deteriorating vision that afflicted his right eye and caused him to squint is barely noticeable in this ca. 1756 portrait of the great Swiss mathematician Leonhard Euler.

match the clever rhetoric of the philosophers in the king's employ, Euler was considered by Frederick to be ultimately too "simple" for his tastes. Meanwhile, the vision in Euler's right eye deteriorated so much that he seemed to squint perpetually, earning him the unflattering moniker Cyclops.

Euler eventually returned to his adoptive Russia in 1766 at the warm invitation of Catherine II (Catherine the Great), who was eager to have the greatest mathematician of the time back in her country. Catherine II had become empress in 1762 after the assassination of her husband, Peter III. Four years later Euler—then fifty-nine years of age—was back at his old academy. Through an operation, doctors attempted to restore the sight in his left eye, which was clouded by a cataract, but to Euler's despair infection soon set in and his sight deteriorated again. The empress provided a large furnished house for the mathematician and his dependents and even assigned one of her cooks to make all their meals.[7]

As blindness set in, Euler had to make accommodations. He dictated equations to his grown sons, and they wrote them out on a board. His faith kept him in good spirits while his eyesight deteriorated, and his mental abilities were so immensely powerful that he performed calculations in his head that others found difficult to solve on paper. The Marquis de Condorcet, a French mathematician with whom Euler had worked, recounted that when Euler was already blind, two of his students summed up seventeen terms of a convergent series and achieved results that disagreed, according to this story, in the fiftieth decimal place. Asked to decide who was right, Euler performed the entire long calculation in his head and determined the correct answer.[8]

Catherine II, a.k.a. Catherine the Great, acted as Euler's patron during his stay in Russia from 1766 until his death in St. Petersburg in 1783.

In 1771 a fire raged in Saint Petersburg, and Euler's house was destroyed. Euler's Swiss servant Grimm courageously carried him out of the house to safety and also rescued his wife; the empress then restored the furniture and the library of her favorite mathematician. Five years later Euler's wife of forty years died. He later married her half sister, Salome Abigail Gsell.

Despite his misfortunes, Euler never lost his faith in God. In fact, his belief in the divine order of mathematics was reflected in his encounter with the French philosopher and famous atheist Denis Diderot (1713–84). Diderot visited the academy and worked hard to convert all the scholars and scientists to atheism. According to Augustus de Morgan (*Budget of Paradoxes*, 1782), Euler leaped at the chance to capitalize on the philosopher's ignorance of mathematics and challenge him. "Diderot was informed that a learned mathematician was in possession of an algebraic demonstration of the existence of God, and would give it before all the Court, if he desired to hear it. Diderot gladly consented . . . Euler advanced toward Diderot, and said gravely, and in a tone of perfect conviction: 'Sir, $(a + b^n)/n = x$, hence God exists—reply!'" Diderot understood nothing about mathematics, so he remained silent, but the wild laughter of everyone around him humiliated him so much that he quickly packed his bags and returned to France.[9]

IT IS HARD TO BELIEVE how a mathematician could have produced as much as Euler had. He was especially interested in calculus, on which he wrote a series of books and research papers. He also wrote elementary books on mathematics for the Russian secondary education system. His research on the calculus of variations created an important new field, and he founded analytical mechanics as well. Euler also initiated the mathematical study of the rotations of rigid bodies and discovered the equations of fluid dynamics used in the field of hydrodynamics. He proposed the letter e for the natural number that is the base of the natural logarithms and proposed the Greek letter π for the ratio of the circumference of the circle to its diameter. We also credit Euler with what is considered the most beautiful equation in mathematics: $e^{i\pi} + 1 = 0$. The equation incorporates basic elements of mathematics, including the essential numbers 0, 1, e, and π, as well as i and the basic elements of arithmetic (the signs for addition, multiplication, and exponentiation, as well as the equal sign).

Euler was a great universalist, contributing to many areas of mathematics with an exceptional intensity. He was most fond of calculations, however, performing them constantly in his mind right up until the moment of his death. On September 18, 1783, Euler spent the afternoon calculating the rate of ascent through the air of a balloon and then dined with his family and friends, discussing the planet Uranus, which had been discovered two years earlier by the German-born British astronomer Sir William Herschel. As an after-dinner exercise, Euler calculated the approximate orbit of this new planet. His grandson then came to play with him and, while playing with the boy and drinking tea, he suffered a stroke, uttered, "I die," and collapsed. In his eulogy for Euler, the Marquis de Condorcet said that at that moment, "Euler ceased to live and calculate."[10]

CARL GAUSS

The German prodigy Carl Friedrich Gauss (1777–1855) was perhaps the most versatile mathematical genius in history. Gauss was born in Brunswick, Germany, to a family of bricklayers and gardeners, and his father was deter-mined that young Carl become a bricklayer as well. A strict disciplinarian, Carl's father was cruel to his children and didn't encourage them to do any-thing except hard, menial labor. He especially discouraged his son's intellec-tual tendencies. Luckily, Gauss's mother had a softer and far more sensitive disposition, and throughout his life Gauss remained attached to her.

Gauss began to show signs of an abnormally high intelligence at age two, so his mother, Dorothea Gauss, constantly asked "experts" to eval-uate his development. Once, before he was three years old, he watched his father, who had by then advanced to a supervisory position, calculate the accounts so that he could pay the bricklayers in his group. Pointing to an amount to be paid, Gauss suddenly said, "This is wrong." The tod-dler had caught his father making an arithmetical error. When the father rechecked his work, he realized that the child was right; but the son's fantastic ability didn't seem to improve the father's view of intellectual pursuits. Fortunately, his wife's younger brother, Friedrich Benz, was an intelligent man who, after becoming aware of the young child's genius, rescued his nephew from a life of hard labor.

Gauss taught himself to read before he had any schooling, and—apparently by watching adults write numbers—he could decipher their meaning and the rules of arithmetic by logical deduction. As an old man, he would joke that he knew how to calculate before he could talk.[11] Very much like Euler, Gauss was a natural calculating machine who would do complicated mathematical computations in his mind as a matter of course.

At age seven, Gauss entered school. Two years later, in a class on arithmetic, a stern schoolmaster named Buettner gave his students a problem that was supposed to be very time-consuming. He asked them to add the long series of numbers from 1 to 100. Gauss already knew, at that age, that he didn't have to add up all the numbers, as his schoolmates were painstakingly doing. Instead, he apparently arranged the numbers 0 through 100 in two rows, one under the other. The top row arranged the numbers going "forward" from left to right; the second row arranged them going "backward" from left to right:

0	1	2	3	4	5	...	98	99	100
100	99	98	97	96	95	...	2	1	0
100	100	100	100	100	100	...	100	100	100

Carl Friedrich Gauss, depicted in this lithograph at the age of fifty, is known for the rigor he brought to the study of mathematics.

By arranging all the numbers from 0 to 100 thus, the young genius could see that the sum of all the pairs was, identically, 100 and that there were 101 such pairs. The total of 101×100 equaled 10,100, and since that number represented twice the sum of the integers from 1 to 100, he divided the sum by 2, which gave him 5,050. Generalizing this procedure gives us the rule for the sum of integers from 1 to any given number, n: $1 + 2 + 3 + ... + n = n(n+1)/2$.

Buettner, known for terrorizing his students, apparently

softened when he witnessed the amazing performance of the youngest student in the class and, as a gift, gave him an expensive new book on mathematics. "He is beyond me," he said of Gauss. "I can teach him nothing more."[12] Buettner's assistant, Johann Martin Bartels, who was only seventeen years old at the time, became a close friend of Gauss. Together they studied infinite series, which brought Gauss to derive a proof of the binomial theorem when the exponent is not an integer. This was the first rigorous proof in his young career.

MATHEMATICAL RIGOR

According to the historian of mathematics Eric T. Bell, it was Gauss alone who imposed on post-Greek mathematics the rigor we see in it today.[13] The question of "rigor" has been key to the nature of mathematics throughout its history. The Babylonians and Egyptians had no interest in proofs, only results. And their solutions—perhaps with some exceptions, such as the Pythagorean triples—had been centered on applications. With their invention of pure mathematics (i.e., mathematics for its own sake), the Greeks also invented rigor. Rigor is the requirement that all statements be proved conclusively, in a way that is absolute (given basic assumptions), and that cannot be denied or destroyed with some counterexample.

Many of the Greek mathematicians followed this rigorous approach. An excellent example is Euclid's rigorous proof regarding prime numbers, which is 2,300 years old. Theorem: There are infinitely many prime numbers. Proof: Assume the theorem is false. Then there must be a largest prime; call it p. Let the product of all the prime numbers, plus 1, be represented by the number $n = 2 \times 3 \times 5 \times 7 \times 11 \times 13 \times \ldots \times p + 1$. Is this number a prime? If it is, then you've just exhibited a prime number greater than p. If n is *not* a prime, then by definition it is divisible by one of the primes 2, 3, 5, 7, 11, 13, $\ldots$, p. Call that particular prime number q. Dividing n by q cannot result in an integer, since you will have the added factor $1/q$, so n is not divisible by any prime number and must therefore be prime. Thus the "contradiction" (*reductio ad absurdum*, as the ancients called it) establishes the truth of the theorem.

Rigor, for the most part, was not a major concern in seventeenth-century mathematics. Gauss changed that. His study of infinite series taught him that if one is not careful with assumptions and derivations, one can obtain absurd results. One such example is the absurd conclusion he obtained when he attempted a careless derivation, getting the statement 0 = 1. In order to avoid obtaining such results, which sometimes occur if one is not careful when dealing with the infinite, Gauss imposed strict standards on his work very early on—a practice that became a staple of modern mathematics.

Johann Bartels, coming from an affluent and well-connected family in Brunswick, introduced his brilliant young friend to the Duke of Brunswick, Carl Wilhelm Ferdinand, in 1792. This encounter changed Gauss's life. His father, as always, was completely uninterested in education and would have done nothing to further his son's, despite pleas from his wife and her brother. In any case, the family was poor and could not afford it. Noblemen supported the arts and culture in their realms, often taking artists, writers, intellectuals, and jesters into their entourages and under their sponsorship, but to offer full support for a schoolboy was unusual. Fortunately for Gauss, the Duke of Brunswick was very taken with the amazing intellectual abilities of the fourteen-year-old and, surprisingly, offered to pay for his entire education.

Gauss attended the Collegium Carolinum (Caroline College) in Brunswick, entering at age fifteen and graduating at age eighteen. Of course, he pursued mathematics on his own because he was so far beyond the pedestrian mathematics taught at the college. When he graduated, the duke continued to support his education at the University of Göttingen starting in 1795. By the time he entered the university, Gauss had already invented the powerful technique we know today as the method of least squares, one of the most important principles in mathematical statistics. Least squares is a routine that uses calculus to find the minimum possible sum of the squared errors of a statistical tool for analyzing data.

For example, in order to find the best straight line to describe the movement of a set of points representing observations on two variables, we use Gauss's method of least squares. The routine is used everywhere in theoretical statistics, and his analysis of how errors are actually distributed later led Gauss to derive the Gaussian law of error, also known as the Gaussian distribution. More commonly, we refer to it as the normal distribution, or bell curve.

Given that he had already made significant contributions to mathematics, it is surprising that Gauss had not yet made up his mind on what to study. He was still attracted to language and considered concentrating on philology. Only in March of the following year did he decide to major in mathematics. In 1796 he began to keep a mathematical diary, in which he wrote until 1814. The diary contains only nineteen pages, but those nineteen pages exhibit 146 concisely stated key results that the young mathematician derived in a number of mathematical fields, including number theory, elliptic functions, analysis, and other areas. In 1898, when a grandson of Gauss first lent the diary to the Royal Society of Göttingen for analysis, it revealed that Gauss placed a priority on achieving as many important discoveries in mathematics as he possibly could. Like Newton before him, Gauss seemed to be reluctant, for some reason, to publish many of his great findings, so his discoveries came to light only after his death. One example of a cryptic result is Gauss's July 10, 1796, entry, completed just as he finished his first year at the university.[14] The sentence reads:

$$\text{EUREKA! num} = \Delta + \Delta + \Delta$$

What does this mean? Gauss had discovered that every positive integer is the sum of, at most, three triangular numbers. Recall the triangular numbers of Pythagoras (T_n), each of which is obtained by adding, to the previous number (T_{n-1}), a number greater by one than the difference between the previous number and the one before it $(T_{n-1} - T_{n-2} + 1)$. Thus, in the triangular sequence 1, 3, 6, 10, 15, 21, 28, 36, 45, . . . , the next number is 55. We can test Gauss's finding with examples—29 = 28 + 1; 40 = 36 + 3 + 1—but it is very possible that Gauss, the great rigorous mathematician, had a proof of this relationship that never came to light.

Historian Eric T. Bell quotes Gauss as saying in old age that he had never published the findings in his notebook because, as a young man, he was deluged by so many ideas and had barely any time to record them. Besides, he didn't care whether other people saw them or not; he was pursuing theorems for his own amusement. Apparently, the outcomes resulted from weeks of research, so he likely possessed proofs, but rather than bothering to write them down, he was always advancing to the next great mathematical puzzle. Gauss preferred to revise assiduously any mathematical result he did publish, provide careful proofs, and make sure every detail was in place. His published works were "few, but ripe."[15] While studying at Göttingen, Gauss wrote a book called *Disquisitiones Arithmeticae* (Arithmetical Investigations) about number theory, dedicating the book to his patron, the Duke of Brunswick. The book was a masterpiece. Before he had reached the age of twenty-one, he had completed what would be considered one of the most important mathematics books ever written.

At the university, Gauss also began a lifelong friendship with a Hungarian student named Wolfgang Bolyai, whom he described as "the rarest spirit I ever knew."[16] Since the age of twelve, Gauss had been obsessed with Euclid's geometry—especially the fifth postulate. He tried to prove the fifth postulate based on the prior four, as many had attempted to do before him, but eventually came to the conclusion that this was not possible. His friend Bolyai was trying to do the same thing when Gauss informed him that he knew that attempts at a proof were futile and that, in fact, geometries other than Euclid's were possible!

Through his good friend's son, Janos Bolyai (1802–60), born six years later, Gauss's idea would eventually lead to a generalization of Euclid's work to what we now call non-Euclidean geometries.[17]

THERE ARE DIFFERENT KINDS of non-Euclidean geometries. In elliptic geometry it is acceptable for "parallel" lines to meet. For example, two longitudinal lines of the earth are "parallel" but meet at the poles. In hyperbolic geometry there could be infinitely many parallels to a single line, as on a saddle. The picture at the top of page 153 illustrates both elliptic and hyperbolic geometries, as contrasted with Euclidean.

Notice that on the sphere, triangles are "fat"—i.e., the sum of their angles is more than 180 degrees, the sum of angles of a triangle on Euclid's plane. On a saddle, however, triangles are "thin"—i.e., the sum

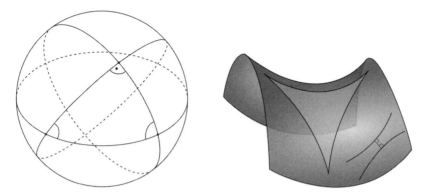

In elliptical geometry (illustrated by the sphere at left), it is possible for two "parallel" lines to meet; in hyperbolic geometry (illustrated by the saddlelike shape at right), the sum of the angles of a triangle is less than 180 degrees.

of their angles is less than 180 degrees. A circle on the sphere has ratio of circumference to diameter less than π, and on a saddle that ratio is greater than π, as can be seen by imagining a circle, and its diameter, placed on each of the geometrical objects in the picture above. These geometries are real and mathematically consistent. Had Euclid lived in a more varied terrain than the flat plains of Egypt, perhaps the curvature of the earth would have led him to adapt his own geometry to realms of more interesting curvature—and perhaps to discard his fifth postulate.

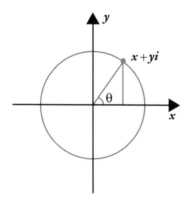

In Gauss's complex plane, the real part of a complex number is represented by a displacement along the x axis, and the imaginary part by a displacement along the y axis.

In 1799, when he was twenty-two, Gauss finished his doctoral dissertation at the University of Helstedt. His dissertation was the first rigorous proof of the Fundamental Theorem of Algebra.[18] Gauss proved that all the roots of an algebraic equation are what we now call complex numbers—numbers of the form $a + bi$, where i is the square root of -1. He was the first mathematician to propose the interpretation of complex numbers as points on the complex plane, thus giving them a precise geometrical meaning.

The complex field is the smallest field in which all quadratic equations have their solutions. (Solutions of quadratic equations can be real, imaginary, or a mixture of both, so they all "live" on the complex plane.) Over his lifetime, Gauss provided four different proofs of the Fundamental Theorem of Algebra, the last one when he was seventy years old.[19] He also launched the field of differential geometry.

Gauss was always fascinated by numbers and their properties—studied in a field we now call number theory, which was in large part launched by his *Disquisitiones Arithmeticae*. He asked himself about the properties of what we call Gaussian integers—that is, complex numbers of the form $a + bi$, where a and b are integers. These numbers have interesting properties. For example, the number 5, which is prime in the "normal" system of numbers, is no longer a prime number in the integral domain of the Gaussian integers. Why? Because it is now factorable into two "prime numbers," the Gaussian integers $1 + 2i$ and $1 - 2i$: $(1 + 2i)(1 - 2i) = 1 + 2i - 2i - 4i^2 = 1 - 4(-1) = 5$.

THE DENSITY OF PRIMES

Returning to "normal" integers, one of the main concerns of number theory is the study of prime numbers. We have seen earlier that Euclid proved that the prime numbers are unending—there is no possible "largest prime number," no matter how far out in the sequence of positive whole numbers you go. Gauss knew this, but he wondered what happened to the frequency of prime numbers as he surveyed the numbers further and further. Did they become more frequent, or less so?

Recall the Sieve of Eratosthenes we saw earlier in the book. As we go from 1 to 100, do the primes become denser (more primes per length of a sequence of numbers) or less dense (fewer prime numbers per length of a sequence of numbers)? Let's take a look (the prime numbers appear against a gray background):

1 2 3 4 5 6 7 8 9 10 11 12 13 14 15 16 17 18 19 20 21

22 23 24 25 26 27 28 29 30 31 32 33 34 35 36 37

38 39 40 41 42 43 44 45 46 47 48 49 50 51 52 53

54 55 56 57 58 59 60 61 62 63 64 65 66 67 68 69

70 71 72 73 74 75 76 77 78 79 80 81 82 83 84 85

86 87 88 89 90 91 92 93 94 95 96 97 98 99 100

Between 1 and 20 there are eight primes; in the next set of twenty positive integers, there are only four primes; in the next set of twenty, there are five primes; the next set of twenty has five primes; and the last set, from 80 to 100, has only three prime numbers. In the first 100 numbers, we can see that there is a moderate decrease in density: but does this trend continue? Might there be 60 primes between 1,900 and 2,000, as opposed to the mere 25 within the first 100 numbers? And if the density does continue to decrease, how fast does it decrease?

Gauss made the first attempt at an answer to this question, one of the most important questions in number theory. On the back of the page of a table of logarithms—he studied Napier's invention with much interest—Gauss wrote in German:

Primzahlen unter a (= ∞) a / la

Recognizing that "l" stands for the natural logarithm, and hence that la is the natural logarithm of a, Gauss's statement is what we call the *prime number theorem*: The number of prime numbers (*primzahlen*) less than any given number, a, approaches the limit a / lna as the number a approaches infinity. This is a relatively slow rate of decrease, but it suggests a constant overall decrease in the number of primes as we approach infinity. It implies that for a large number, a, the probability that it is prime is roughly 1 / lna. The celebrated Riemann hypothesis, named after the German mathematician G. F. B. Riemann, implies bounds on the errors of the estimates provided by the prime number theorem, but to this day the conjecture is still unproved.

In 1805 Gauss married a young woman from Brunswick named Johanna Osthoff. They had three children, but she died giving birth to the third in 1809. Gauss was heartbroken, but for the sake of his children, he remarried the following year and had three children with his second wife, Minna Waldeck.

Gauss didn't get along very well with his children—he was, perhaps, too busy doing mathematics. The exception was Joseph, his son from his first wife, who was fairly gifted in mathematics. Two sons ran away from home as adolescents, immigrated to the United States, and became farmers in Missouri. Later one of them became a wealthy merchant in Saint Louis, which was a major trade port during the colorful days of the Mississippi riverboats. The sons had several children each, and it is believed that Gauss has many descendants living in America today (perhaps excelling as mathematicians).[20]

Meanwhile, when he was seventy years old, the Duke of Brunswick, who had generously supported Gauss, was called to help Germany fight against Napoleon. He was sent to Saint Petersburg to try to enlist Russia's help, but the tsar refused, and the duke was put at the head of a large German force opposing Napoleon's progress at Jena. Gauss was living in a house by the main highway going into Brunswick, and one fall day he saw a wagon entering town carrying the dying duke, who had been critically wounded in battle. Gauss's anguish was palpable—the kind duke had been like a father to him—but after the duke's death another German philanthropist, Alexander von Humboldt, stepped in to support Germany's greatest mathematician. Through Humboldt's patronage Gauss was made the director of the Göttingen Observatory, a position that also allowed him to teach and continue research in mathematics.

Gauss was a great facilitator, encouraging and supporting the work of younger mathematicians. His last student was the brilliant German mathematician Peter Gustav Lejeune Dirichlet (1805–59), who always carried Gauss's *Disquisitiones* with him, even sleeping with the book under his pillow. The great French mathematicians Lagrange and Laplace, who did important work in mathematics as applied to physics and astronomy, were also big fans of Gauss when he was still a young man.

Thus the geniuses that followed the invention of the calculus achieved great breakthroughs in the history of mathematics and, in many ways, transformed the field into a cohesive and rigorous discipline. Collectively, their work affected many areas, including analysis, probability theory, topology, number theory, and geometry. They laid an important foundation for the future of mathematics.

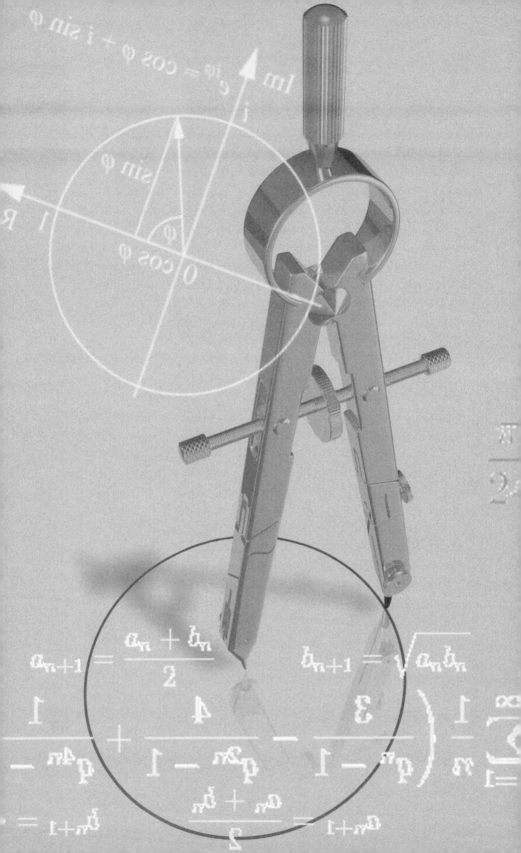

$$a_{n+1} = \frac{a_n + b_n}{2} \qquad b_{n+1} = \sqrt{}$$

PART

V

UPHEAVAL
IN FRANCE

$$e^{i\varphi} = \cos\varphi + i\sin\varphi$$

$$\sin\varphi$$

$$1 \quad Re$$

$$0 \quad \cos\varphi$$

$$\sum_{n=1}^{\infty} \frac{1}{n} \left(\frac{a_n + b_n}{q^n - 1} - \frac{}{q^{2n} - 1} + \right.$$

$$a_{n+1} = \frac{a_n + b_n}{2}$$

$$a_0 = 1 \qquad b_0 = \frac{1}{\sqrt{2}} \qquad t_0 = \frac{1}{4}$$

NAPOLEON'S MATHEMATICIANS

rance has an unusual history in general, and mathematics history in France is no exception. At first the monarchs—members of the so-called *ancien régime*—supported science and mathematics at least as much as the absolute rulers of countries such as Russia and Germany (Prussia). The French Royal Academy of Sciences was founded by a group of scientists and scholars who first met in King Louis XIV's own library in 1666. Six years earlier, in 1660, the Royal Society had been founded in London under a royal charter from Charles II. With Louis XIV's blessing and encouragement, the French scholars and scientists continued to meet in his official library for many years until the

Under the reign of Napoleon Bonaparte, many mathematicians flourished—including Adrien-Marie Legendre, Joseph Lagrange, and Joseph Fourier. Napoleon is depicted in this 1805 portrait by Andrea Appiani.

construction of the Institut de France, with its gilded dome, just across the Seine from the Louvre.

But the French revolution changed all that. In 1793 the National Convention—the political body the revolutionists had elected to rule France instead of its toppled (and guillotined) monarchs—abolished the Royal Academy of Sciences. It was reinstituted under a different, non-Royal guise a couple of years later, but science was not a major concern for the revolutionists. They were more interested in overthrowing everything that had come before them, and the idea of "natural law" was clearly anathema to them because it would have

This 1671 engraving depicts King Louis XIV touring the French Royal Academy of Sciences, which was founded in Paris by a group of scholars in 1666.

supported the old status quo, even though the legitimacy of the monarchy was based on religious belief (God anoints the kings of France through a cleric, such as Saint Rémy, who baptized Clovis, the first king of France) rather than science. Even the names of the months were changed during the Revolution, based upon natural weather patterns in temperate France at different times of the year. For example, the months of Brumaire and Thermidor, roughly covering the same range of dates as November and August in the Gregorian calendar, took their names, respectively, from the Latin words for "fog" and "summer heat."

There is another important consideration that affects our story. In the eighteenth century, the study of science was sometimes considered a luxury. One of the founders of quantum mechanics, Louis de Broglie, was a prince. The French naturalist and mathematician who partially developed the theory of probability, Georges-Louis Buffon, was a count. And we've already met the Marquis de l'Hôpital, another nobleman. In fact, there were many nobles engaged in scientific pursuits—not only in France but throughout Europe. During the Revolution thousands of members of the French nobility were executed, and science and mathematics therefore suffered during this tumultuous and brutal period. The mathematician M. J. Condorcet, a marquis, lost his life despite the fact that his ideas were very much in the direction of the abolition of the monarchy and support for the rights of man. And for those who remained unmolested, how easy could it have been to find the tranquillity and garner the powers of concentration required for conducting mathematical or scientific research while food is scarce and one lives in constant fear of the dreaded knock on the door?

But the Revolution did accomplish one very important achievement in mathematics as applied to society: it established the metric system. This is today a very widely used system around the world—and it is certainly important in the sciences. Ironically, the metric system owed its birth to the revolutionaries' insistence on breaking from the past, as they were doing with the calendar.

THE BIRTH OF THE METER

Adrien-Marie Legendre (1752–1833) was one of the French mathematicians of the revolutionary period, advancing a number of areas of mathematics, including the least-squares method in statistics that was later completed by Gauss. He also helped measure the Paris Meridian, which passes through the center of the Paris Observatory, using triangulation methods. *Legendre polynomials* are polynomials used in mathematical applications such as physics; they result as solutions to a differential equation. Nevertheless, Legendre was left out of the committee of mathematicians who were charged with creating the new system of measures.

In 1790 the French diplomat Talleyrand put forward the idea that as part of the overthrow of everything that came before them, the revolutionaries must choose a new system of weights and measures. The Academy of Sciences was still in existence at the time—it would be abolished three years later (and would be reinstituted still later)—so the problem of setting the new system was referred to a committee of the academy that included the two mathematicians Joseph-Louis Lagrange (1736-1813) and Lazare Carnot (1753-1823). The committee entertained two main proposals after various ideas had been whittled down. One system of measures was to be based on the decimal (base-10) number system, and the other on the duodecimal (base-12) number system.

Lagrange was greatly opposed to the duodecimal system, which was gaining strong support in the committee. The arguments for a base-12 system were rooted in the fact that 12 is divisible by more numbers than 10, so presumably it would be easier to calculate in this system. (The divisors of 12 are 1, 2, 3, 4, 6, and 12, while 10 is divisible only by 1, 2, 5, and 10.) Lagrange reportedly led a charge in the *opposite* direction: "Why should the base of the system of measurements we adopt be a number divisible by many numbers?" he asked. "The system should be based on a *prime* number, such as 11," he argued. However, it has been conjectured that Lagrange led with this argument only to derail the supporters of the duodecimal system.[1] Eventually, Legendre's celebrated success in measuring the meridian was brought into account, and since that measurement was based on the decimal system, the decimalists won. The *meter* was thus defined, based on the meridian, as one ten-millionth of the distance between the equator and the poles.

Recall Eratosthenes' estimation of the circumference of the earth as 250,000 stades. A stade is about a tenth of a mile, making that circumference 25,000 miles. This measurement is equivalent to about 40,000 kilometers, and a quarter of this length is the distance from the equator to the pole. One ten-millionth of 10,000 kilometers is 1 meter. Lagrange and Carnot—a mathematician and military leader who worked in geometry and calculus—had won! Years later it would be none other than Carnot who would promote a young colonel named Napoleon Bonaparte to the rank of general, launching a career that would change the world. But Lagrange was a far more important mathematician than Carnot.

JOSEPH LAGRANGE

In the words of Napoleon, "Lagrange is the lofty pyramid of the mathematical sciences." The French emperor would eventually appoint the brilliant mathematician to the positions of senator, Count of the Empire, and Grand Officer of the Legion of Honor, the honorary body established by Napoleon.[2]

Joseph-Louis Lagrange was born in Turin, Italy, to French-Italian parents. His grandfather was a captain of the French cavalry and offered his service to the king of Sardinia before settling in Turin and marrying a daughter of the influential Conti family. Their son became the treasurer of war for Sardinia and married Marie-Therese Gros, the daughter of a wealthy physician. The couple had eleven children, but only the youngest—Joseph-Louis, born in 1736—survived to adulthood.

Joseph's father was a speculator, and although he had inherited money from his own parents and had married into a family of means, he lost most of his assets. As an old man, Joseph once commented, "If I had inherited a fortune I should probably not have cast my lot with mathematics."[3] Mathematics would have been the poorer, for Lagrange contributed much during his mathematical career.

At school Lagrange was interested in the classics, and only through classical studies became familiar with the works of the ancient Greek geometers. He was not impressed with geometry, however, and would have become a classical scholar had he not come upon a book by Edmond Halley, Newton's friend, which explained Newton's calculus with great enthusiasm. Lagrange studied Halley's work on his own and became enamored with the new mathematics of the time. It is said that he was only sixteen years old when he became a professor of mathematics at the Royal School of Artillery in Turin, although that claim has been disputed.[4]

At any rate, when he was nineteen, Lagrange wrote a book called *Mécanique Analytique* (*Analytical Mechanics*), which was recognized as one of the world's greatest books on mathematics when it was published thirty-three years later. His dislike of geometry—ancient Greek or otherwise—was evident in the book. The preface stated, "No diagrams will be found in this work."[5] Instead, Lagrange claimed that all of mechanics could be understood as a generalized geometry that is four-dimensional—including three spatial dimensions, plus the dimension of time—and therefore cannot benefit from diagrams or drawings of any kind. With this deep insight, he anticipated Einstein, whose theories of relativity wed space and time to create space-time, and freed mechanics from its Greek origins. Instead, Lagrange introduced purely analytical methods, which utilized algebraic tools instead of geometrical ones.

The teenage professor in Turin was the main force behind the founding of the Turin Academy of Sciences, and he contributed to its proceedings, encouraging his students—who were older than himself—to conduct research in mathematics, which would subsequently be published by the academy. One of his own contributions was the development of the calculus of variations. His work elaborated on the achievements of Newton and founded mechanics on a solid mathematical base.

In his early twenties, Lagrange contributed to the theory of probability by applying calculus to statistical problems. He also worked on the mathematical theory of sound by considering the movement of air particles in elastic settings and studied the mathematical theory of vibrating strings. At age twenty-three he was recognized all over the world as a mathematician as great as Euler and the Bernoullis.[6]

Lagrange sent some of his work to Euler, who was then in Berlin. Euler, who had been working on similar problems with little success, quickly recognized that the young French-Italian mathematician's work would enable him to solve some of these problems. Being kind and fair, he encouraged Lagrange to publish his work first and only afterward arranged for publication of his own papers based on Lagrange's breakthrough in the calculus of variation. He didn't want there to be any doubt about who was first to develop the new methodology.

Euler then devised a plan to get Lagrange to join him at the Royal Prussian Academy. At the insistence of Euler and other mathematicians, Frederick the Great invited Lagrange to Berlin. Before going to Berlin, Lagrange won a prize from the French Academy of Sciences for solving the problem of the moon's libration, which explains why, with small variations, we see only one face of the moon. Lagrange made strides in our understanding of the celebrated "three-body problem," in which the sun, earth, and moon exert gravitational forces on each other. He later won several more prizes from the academy for partial solutions to this problem, which is very complicated. In 1766, when he was thirty,

Lagrange was finally welcomed to the Berlin Academy and, after a short while, took Euler's position as director of mathematics. He worked for years at the academy revising his *Mécanique Analytique* masterpiece with the help of Adrien-Marie Legendre.

Lagrange made contributions to number theory, but he did even more important work in algebra, in the theory of equations. The Babylonians, Egyptians, and Greeks, followed by the Arabs, Italians, and others, had done extensive work on solving equations of increasing order: linear (where the unknown, x, is in the first power), quadratic (second power), cubic (third power), and quartic (fourth power). But Lagrange sought a general method of solution for equations of any order. He began to understand that the number of different orders in which solutions, expressed in terms of the coefficients of an equation, can be arranged had something to do with the solvability of an equation. Nevertheless, he did not find a general method, although his pioneering investigations would be carried to a far greater extent by the young Frenchman Évariste Galois in the 1800s.

In mechanics Lagrange invented a function we now call the Lagrangian, in his honor. This function lists and ties together all the relevant parameters of a physical situation. Today it is of immense importance in modern physics—particularly quantum mechanics and modern particle theory, in addition to the Newtonian mechanics for which Lagrange had invented it in the first place. A brilliant twentieth-century mathematician, Emmy Noether, described later in this book, used the Lagrangian function—in addition to the work of Galois and the Norwegian mathematician Lie, who extended Lagrange's work on equations—in a study of symmetry that established conservation laws in physics. These laws govern such physical

Among mathematician Joseph Lagrange's many contributions to the field is his groundbreaking book *Mécanique Analytique*, published in 1788—thirty-three years after it was written.

processes as the conservation of energy, momentum, electric charge, and other parameters used in modern physics.

The Lagrange multiplier, which allows one to find maximum and minimum points of functions given some constraints, is another important discovery in applied mathematics that extends calculus ideas to far more general and complicated situations than previously achieved. Lagrange also pioneered the use of the determinant, which he used to find the areas of triangles and the volumes of tetrahedra.[7]

When Frederick the Great died in 1786, a rise in nationalism that had been kept at bay by the monarch brought pressure on foreign-born members of his academy to leave Germany. Lagrange was a favorite of the academy, however, so when he resigned he promised to keep sending papers to be published in the academy's proceedings. From the Royal Prussian Academy in Berlin, Lagrange moved to the Royal Academy of Sciences in Paris at the invitation of Louis XVI himself. The king's wife, Marie Antoinette, admired him as well, and he lived for a while in the Louvre as special guest of the royal family.

Lagrange was now feeling his age—he was in his fifties—and the move to Paris brought on a deep depression, which Marie Antoinette tried to ameliorate with lively conversations. Lagrange spoke little at the lavish parties he was invited to and was described as staring blankly out the window with his back to the other guests.[8] He also lost his interest in mathematics. The only high point during this dark period was his friendship with the great French chemist Antoine Lavoisier (1743–94). Lagrange had begun to believe that the new age belonged to chemistry and other sciences and that mathematics had lost its glitter. When told of a fellow mathematician's great discovery, he answered, "All the better; I began it; I won't have to finish it."[9]

When the French Revolution began, he was warned that it was in his best interest to return to Berlin, where he was always welcome, to avoid the dangers at home. Lagrange refused to leave France, and felt no sympathy for either the Royalists or the Revolutionists, but when the Terror ensued and Lavoisier was guillotined for trumped-up charges, Lagrange became angry and even more depressed. Despite his disillusionment with mathematics and with life in general, he was appointed president of the committee that determined the new weights and measures system, and it was his personal triumph when it was decided that the system would be metric rather than duodecimal.

Among those who were guillotined during France's 1793–94 Reign of Terror were chemist Antoine Lavoisier, who helped Lagrange avoid persecution by the authorities, and King Louis XVI—the only king of France ever to be executed.

Thanks to a young woman almost forty years his junior, Lagrange's depression finally lifted. Renée-Françoise-Adélaide Le Monnier—the daughter of an astronomer friend, Pierre-Charles Le Monnier (1715–99)—met Lagrange when he was fifty-six years old and, feeling sorry for the depressed genius, offered to marry him. They wed, and his life changed overnight. He was so much in love with his very young wife that he couldn't bear to be away from her—he accompanied her to balls and banquets he would have done anything he could to avoid prior to their marriage.

When Napoleon came into prominence, he would often talk to Lagrange about mathematics and its place in society whenever the general was in Paris between military campaigns, and when Napoleon became emperor, he bestowed on Lagrange many high honors. By the time Lagrange had completed the last revision of *Mécanique Analytique* for a second edition, he had begun to suffer from dizzy spells. His body had finally gotten tired, and he died in 1813 at the age of seventy-five.

PIERRE-SIMON LAPLACE

Lagrange's fellow mathematician on the weights and measures committee that gave the world the metric system was the mathematician Pierre-Simon de Laplace (1749–1827). A mathematical astronomer, Laplace was often called the Newton of France.[10] Laplace's origins are shrouded in mystery, due in part to the fact that he was born poor and tried hard to hide where he came from once he obtained noble status. We know he was born in the Calvados region of Normandy, in northwest France. He started off as a theologian but later in life turned atheist. As a young man, he absorbed mathematics quickly and, with strong recommendations from his teachers, left for the big city: Paris.

In the French capital, Laplace tried to get an audience with Jean Le Rond d'Alembert (1717–83), a successful mathematician of equally humble origins. D'Alembert was the illegitimate son of Louis-Camus Destouches, a French artillery officer, and Claudine Guérin de Tencin, a well-known writer, former nun, and sister of a cardinal. As his father was abroad when the baby was born, Madame de Tencin decided she would just as well get rid of the unwanted child by placing him on the steps of the chapel of Saint Jean le Rond—where he was baptized and from which he took his name—near the Cathedral of Notre Dame in Paris. He was then taken to an orphanage, and when his father returned to France, he arranged for him to be adopted by a glassworker named Rousseau and his wife.

Though Chevalier Destouches kept his paternity a secret, he helped pay for his son's upbringing and education. After he died, when Jean was nine, his family continued to support the officer's illegitimate son. When Jean enrolled at the Collège des Quatre Nations, he assumed the fake name Jean-Baptiste Daremberg, since he could not use his father's name and did not want to use the name of his adoptive parents. He later changed it to something slightly more noble

The mathematician Jean Le Rond d'Alembert, among his many other achievements, wrote more than one thousand articles for Denis Diderot's famous *Encyclopédie*, published between 1751 and 1772. He also served as coeditor of the publication.

sounding: Jean Le Rond d'Alembert. Meanwhile, d'Alembert's birth mother found out who had taken her baby, and when it became clear that he was brilliant and on his way to becoming a major mathematician, she tried to get him back. "You are only my stepmother," the young man told his wealthy and aristocratic biological mother, insisting that the poor couple who had been raising him were his real parents.

Having made it in life, d'Alembert faced Laplace, a young man from the provinces who dropped several glowing letters of recommendation on his desk. D'Alembert read the letters but seemed unimpressed, and the young mathematician left. A few days later, trying again to gain an entry to Paris academia through d'Alembert, he sent him a short paper he had written on the mathematical principles of mechanics, the area studied so assiduously by Lagrange. Some time later a letter arrived back from d'Alembert. "Sir," he wrote, "you see that I paid little enough attention to your recommendations; you don't need any. You have introduced yourself better. That is enough for me; my support is your due."[11] A few days later, on d'Alembert's recommendation, Laplace was appointed professor of mathematics at the École Militaire.

Laplace began his research while teaching at that school, laying the groundwork for his world-famous masterpiece, *Mécanique Céleste* (*Celestial Mechanics*). He decided to apply Newton's physics and the mathematics of his theory of universal gravitation to a study of the entire solar system, with all its known planets.

It was a monumental work but borrowed heavily from Newtonian principles, which Laplace acknowledged. However, the book also relied on the work of Lagrange—most significantly, the idea of a gravitational potential, which is a key concept in the physics of gravitation today—and methods of analysis developed by Legendre. Laplace did not reference or acknowledge the latter two mathematicians.[12]

Pierre-Simon Laplace's five-volume masterwork, *Mécanique Céleste*, was published between 1799 and 1825.

In *Mécanique Céleste*, Laplace attacked the problem of how bodies mutually affect each other gravitationally. This grand extension of the work of Lagrange to the entire system of planets and the sun involved many questions, including: Could one of the planets veer off into space and leave our solar system? Could Mercury slip from its orbit and fall into the sun? Could the moon begin to move away from Earth and crash into Mars? All these questions comprise one overarching riddle: Is our solar system stable, and if so, why? Laplace addressed this great set of questions by analyzing the gravitational forces acting between and among the bodies in the entire known solar system. In the end he was able to prove that the solar system was stable.

Mécanique Céleste went far beyond Newton in its mathematical study of gravitation. Newton had explained the force of universal gravitation but assumed that the solar system was held together by divine intervention, thus reconciling his deep religious beliefs with the scientific and mathematical principles he had discovered.

Thanks to his great achievement, at the age of twenty-four Laplace was inducted into the French Academy of Sciences. He spent his entire life perfecting his theory of mathematical astronomy, but he also did work in probability theory, as applied to astronomy. He is known for inventing what we call the Laplacian operator, given by the sum of the second-order partial derivatives of a function, which is very useful in physics and other applications of mathematics.

WHEN THE REVOLUTION OCCURRED, Laplace and Lagrange are said to have escaped the guillotine because they were expedient in their abilities to calculate trajectories of artillery shells. The mathematician M. J. Condorcet fared much worse. In jail, taken in by a sweep of all nonworkers and aristocrats, he asked for an omelet. A nobleman, Condorcet had never seen an omelet being cooked. "How many eggs do you want in your omelet?" his jailers asked. He thought about it for a minute and then said, "Twelve." The jailers then asked to see his hands, and when they confirmed to their satisfaction that the man probably never did any hard manual work in his life, they sent him to be guillotined.

When the Terror passed, Laplace decided to become a politician. Having attained high status through his achievements and membership in the prestigious French Academy, the great mathematical astronomer decided to try to obtain real power. He was somewhat mercurial, however, changing from Republican to Royalist whenever the political winds shifted. When Napoleon came to power, he heaped honors on Laplace, who as a scientist had brought glory to France. In addition to receiving the Grand Cross of the Legion of Honor and assuming the title of Count of the Empire, the humbly born mathematician eventually became a marquis.

Laplace made a gift to Napoleon of his *Mécanique Céleste*. Perhaps to tease him, Napoleon commented, "You have written this huge book on the system of the world without once mentioning the maker of the universe." Laplace answered, "Sire, I had no need for that hypothesis." Napoleon later repeated his exchange with Laplace to his other mathematician friend, Joseph Lagrange. The latter, ever tactful and diplomatic, responded, "Ah, sire, but it's such a beautiful hypothesis."

When Napoleon was defeated at Waterloo, it was Laplace—now an official in the French government—who had to sign the decree banishing Napoleon to Saint Helena. There the fallen emperor summed up what he thought of the mathematician-turned-politician: "A mathematician of the first rank, Laplace quickly revealed himself as only a mediocre administrator; from his first work we saw that we had been deceived."[13]

JOSEPH FOURIER

During his reign, Napoleon enjoyed the friendship of several other French mathematicians. One of the most important among these was Joseph Fourier (1768–1830). Fourier was the son of a tailor in Auxerre. When he was young, both his parents died, so he was brought to the care of the bishop of Auxerre, who arranged for him to be adopted and admitted to the local military college.

Fourier was a troubled boy who refused to listen to teachers, and he nearly ended up on the streets, but then he discovered mathematics, which kept him fascinated and occupied until graduation. His low social status squashed his chances of becoming a soldier, and the French Revolution prevented him from entering the priesthood, so he concentrated on studying mathematics.

At age twenty-one Fourier arrived in Paris and presented to the French Academy of Sciences his work on numerical solutions of equations used in physics. He was involved in revolutionary politics but abhorred the Terror that followed. He then studied at the École Normale Supérieure in Paris and excelled. One of his favorite professors was Lagrange, and he took courses from Laplace as well.

His early success presenting a paper to the academy, his studies of mathematics at the École Normale Supérieure, and his connections with Lagrange and Laplace helped him to attain a position at the prestigious École Polytechnique in Paris. There he met Napoleon, who was seeking mathematical help in determining artillery trajectories while planning his early campaigns. When Napoleon went on his conquest of Egypt in 1798, he took Fourier with him.

In his attempt to "liberate" the Egyptians from their "uncultured" state, Napoleon founded the Egyptian Institute, a Cairo-based offshoot of the Institut de France and its associated Academy of Sciences. As cofounder and secretary of the institute, Fourier was involved in archaeological excavations carried out by its staff. The tides of war eventually turned against the French, however, and in 1799 Napoleon felt compelled to depart, leaving Fourier behind to help administer the territory.

Fourier returned to France only three years later, in 1801. After resuming his post at the École Polytechnique in Paris, Napoleon asked him to take an administrative post headquartered at Grenoble, in Isère, the prefect of which had just died. There he created his masterpiece on the mathematical study of heat. This beautiful piece of applied mathematics describes how heat is conducted, using differential equations, an element of advanced calculus. In particular, Fourier simplified the study of heat conduction by representing functions as trigonometric series—a method we refer to as Fourier analysis.

Fourier found a way of deconstructing a set of data into its frequencies (as in music) and analyzing it. The new techniques Fourier developed in mathematics are extremely important today. Similar methods are used in economic and stock-market analyses, and Fourier's derivation of the heat equation later led to other advances in physics, including methods in quantum mechanics.

At Grenoble, Fourier discovered the remains of his great-uncle, Pierre Fourier, who had been canonized. When Napoleon escaped from his imprisonment on Elba in 1815, he passed with his followers through Grenoble and came to see his old friend. When he did not find him, the deposed emperor suspected that, in his absence, the man he had once made a top administrator in Egypt had switched alliances and gone on to support the Bourbons, the traditional rulers of France. Nevertheless, he appointed Fourier to the prefecture of Rhône. Napoleon planned an offensive against the British and Prussian armies, who hoped to unseat him from power. Fourier warned him that his plan would not succeed, but Napoleon ignored the advice of his prized mathematician. Waterloo would prove Fourier right.

Fourier became the Permanent Secretary of the Academy of Sciences and never missed an opportunity to regale his fellow scientists with stories about his adventures with Napoleon in Egypt. The man who explained mathematically how heat is conducted had apparently developed the conviction that the desert heat of Egypt was salutary, so he tried to emulate it in cold and humid France, sleeping in a sweltering bedroom that no one else could survive in for long. Eventually this habit took a toll on his health and he died of heart problems when he was sixty-two.

GASPARD MONGE

Another French mathematician, Gaspard Monge (1746–1818), developed a particularly close relationship with Napoleon, accompanying both him and Fourier on the 1798 conquest of Egypt. Monge was born in Beaune, Burgundy, to a family of peddlers and knife grinders. His father held education in high esteem, however, and sent all his sons to college. But among all the successful Monge sons, Gaspard was the brightest. When he was fourteen he built a fire engine, to the astonishment of everyone around him. A couple of years later he drew a very accurate map of the region where he lived. He later explained that it was his spatial intuition that helped him accomplish both tasks. In fact, this uncanny ability to visualize shapes and forms in three dimensions allowed him to develop descriptive geometry: a way of capturing three-dimensional objects on two-dimensional paper using ingenious graphing techniques that he invented. His ingenuity won him awards at school, and his invention of descriptive geometry earned him an invitation to lecture at the École

Normale Supérieure in Paris, which had recently been founded. One of the people who sat in on his lectures was the great Lagrange, who was fascinated by this new, practical geometry. At the school, he also met Fourier. Eventually, Monge was offered a professorship at the University of Mézières.

Once, at a party, Monge heard a nobleman vulgarly disparage a young widow who had rejected his advances. Monge rushed to defend the unknown lady's honor, punching her insulter in the mouth. At another party later on, he was introduced to the woman whose honor he had jumped to defend, Madame Horbon, and fell for the beautiful young widow of twenty. He proposed to her then and there, and after taking time to put her late husband's affairs in order, she agreed. Monge and Horbon married in 1777. It proved a good match—Monge's young wife even saved his life during the Revolution, while he was in Paris working with d'Alembert and Condorcet. It was a frequent occurrence in times of terror for people to be denounced for various fabricated crimes. One day Monge's wife happened to discover that her husband had been denounced by the Revolutionaries and rushed to Paris to warn him. The couple escaped to the countryside, where it was safe.

In 1792 Monge briefly met Napoleon, who was then a young artillery officer. Four years later, when Napoleon became the general in command of the French army invading Italy, he wrote a letter to Monge thanking him for his kindness when they first met. It was the beginning of a very close, warm friendship.

In 1798, when Napoleon embarked on his naval invasion of Egypt— Leibniz's dream realized more than a century later through Napoleon— he took with him his two most favored mathematicians: Monge and Fourier. The two mathematicians were among a select group of only a dozen people whom Napoleon trusted with his plan of attacking, conquering, and then "civilizing" the Egyptians. A fleet of five hundred French ships left France in the spring of 1798, arriving in Malta on June 9 and taking three days to conquer the island. Here, Monge founded fifteen schools as well as a university modeled after the École Polytechnique. They continued toward Egypt, reaching Alexandria on July 1. Monge was reportedly first in the party to come ashore.

In Egypt, Napoleon asked the mathematician to lead a cavalry regiment in an attack on the fortifications of the Nile, and at one point

During the 1798 Battle of the Pyramids, in which Napoleon's army achieved a decisive victory over Egypt, French mathematician Gaspard Monge served loyally under his commander. He was rewarded handsomely by the emperor, but his honors were taken from him when the monarchy was restored.

Monge was about to be killed by an Egyptian soldier. Anticipating the fatal blow to his favorite mathematician, Napoleon galloped as fast as he could toward the combatants and saved Monge. On July 20 the French won the decisive Battle of the Pyramids, and Napoleon was master of Egypt. After founding the Egyptian Institute, he put Monge in charge of determining which works of art and ancient artifacts would be brought back to France as war booty.[14] So when we visit the vast Egyptology collection in the Louvre, or admire the obelisk at the Place de la Concorde in Paris, we have Monge to thank—or blame—for their presence in France. But the Egyptians apparently didn't want to be "civilized" by the arrogant Napoleon, and massacred many of his troops. On his return to France, the emperor took Monge with him; as we recall, Fourier was to remain there for some time.

THE RESTORATION OF THE MONARCHY in France, following the fall of Napoleon, brought trouble to the mathematicians closest to the emperor. Carnot had to leave France for exile in Magdeburg, Germany, and Monge was stripped of his many honors and robbed of his position at the École Polytechnique. He died soon afterward. Lagrange had died a few years earlier, but Legendre remained politically quiet and continued to publish. Laplace, with his ever-shifting political allegiances, could easily survive under any regime.

The period of unrest that followed the temporary restoration of the French monarchy affected the lives of many mathematicians. One of them was a very young genius whose involvement with anti-Royalist politics would ruin his career and destroy his life at a young age. In his very few active years of life, however, he was able to catapult algebra to the fore and solve key problems that had stumped many able mathematicians over the centuries. We will meet him next.

———— • ◆ • ————

DUEL AT DAWN

T he story of the incredibly brilliant French mathematician Évariste Galois is one of the most romantic
and powerful in history. Ever since the time of
ancient Greece, mathematicians had been looking for ways
of generalizing the method of finding solutions to equations
of increasing order, in terms of their coefficients. For example,
the simple linear (first-order) equation $2x - 4 = 0$ is solved
in terms of its coefficients 2 and 4 as $x = {}^4\!/_2 = 2$. In this
equation, 2 is the coefficient of x to the first power, and
4 is the coefficient of x to the zero power. We all know the
quadratic formula used to solve quadratic equations in terms
of the equation's coefficients. Tartaglia, Cardano, Fior, and
del Ferro have shown that similar formulas can be obtained
for cubic and quartic equations as well. But no such general

Eugène Delacroix's famous painting *Liberty Leading the People* has become a symbol of revolutionary
fervor in France. Mathematician Évariste Galois's father was strongly pro-Revolutionary, and as a young
man, Galois benefited from Napoleon's emphasis on the study of mathematics and science.

formula for solving an equation in terms of its coefficients had been known for the quintic, or fifth-order, equation—or for higher-order equations. The question that mathematicians had hoped to answer was: What is the general formula for solving a fifth-order or higher-order equation? And if no such formula exists, then why?

The answer to this conundrum was provided by the work of Galois, and as a bonus it also led to an understanding of the impossibility of the three classical problems of antiquity— which, as we recall, are squaring the circle, doubling the cube, and trisecting an arbitrary angle. Galois's work also launched group theory, an immensely important part of modern abstract algebra. Incredibly, all these breakthroughs materialized from the work of a very young person who blossomed intellectually between the ages of sixteen and twenty—the age at which he tragically died.

ÉVARISTE GALOIS

Évariste Galois was born on October 25, 1811, in the town of Bourg-la-Reine, a few miles south of Paris. During the French Revolution, the name of the town was changed (*reine* means "queen"). The revolutionaries detested royalty, so they changed the name of the town to Bourg-l'Egalité (town of equality).

Galois's father, Nicolas Gabriel Galois, was a politician, a one-time mayor of the town, and principal of the local school. After Napoleon's demise and the return to power of the Bourbons under Louis Philippe of Orleans, the so-called King of the French, the elder Galois turned fiercely anti-Royalist. His town, however, had many Royalists who constantly attacked him through libelous statements in newspaper articles.

Galois's mother, Adélaide Marie Demante, was the daughter of a judge. She educated her son at home, focusing much of her lessons on the law, as well as philosophy, religion, and literature. At age twelve Galois entered the Royal School of Louis-le-Grand, where he proved to be a brilliant student, but teachers described him as having "bizarre manners" and as being "rebellious." The school was originally mostly concerned with teaching the classics, and mathematics and science were considered nonessential. During Napoleonic times, however, this trend was reversed since, as we know, the emperor favored these disciplines; but the classical tradition was still much in place in the grade Galois was in. He took a course in rhetoric and performed so badly that he was sent back a grade.

Galois's failure in rhetoric was fortuitous, as it turned out. The lower grade was already designed according to the Napoleonic system, so he was allowed to study mathematics. On his own, the young Galois read beyond what was taught in class, including the mathematical works of Gauss, Euler, Lagrange, and Legendre. He was especially interested in Legendre's book *Elements of Geometry,* which he read with ease and excitement when he was only fifteen. But it was the research papers by Lagrange that inspired him, while still a teenager, to try to apply Lagrange's methods to equations that could not, at that time, be solved.

In 1828 Galois attempted the entrance examinations to the prestigious École Polytechnique but failed, so he had to remain at Louis-le-Grand. Luckily for Galois, Louis Richard, the teacher of an advanced mathematics course he took, recognized that he had a genius in the class and kept all Galois's homework assignments.

Galois was a mathematics prodigy at a young age; when he was seventeen, one of his teachers began to preserve his homework assignments, recognizing their future importance.

(After Galois's tragic death, Richard would give these papers to the mathematician Charles Hermite, who later became famous through his own work in various mathematical spheres.) Richard encouraged Galois to publish his early research results. After an article by Galois appeared, on April 1, 1829, in the journal *Annales de Mathématiques,* the seventeen-year-old wrote a more extensive paper and sent it to the French Academy of Sciences, hoping to receive recognition for his important discoveries in algebra from this great academic body.

There, Galois's paper was supposed to be read by the eminent mathematician Augustin-Louis Cauchy (1789–1857), who did groundbreaking work on determinants, a name he coined, as well as complex analysis and other areas. The great Cauchy lost Galois's paper, however. By then, Cauchy had already lost another paper sent him by a young genius working on very similar problems in the theory of equations: Niels Abel.

THE ABEL-RUFFINI THEOREM

Niels Henrik Abel (1802–29) was born in the village of Findö, Norway, to a very large family. His father, the pastor of the village, was also a politician involved in the writing of Norway's new constitution. When he was sixteen, Abel's teacher gave him Gauss's *Disquisitiones Arithmeticae,* hoping the bright student would enjoy it. Not only did Abel enjoy the book, but he was able to fill in details that had been left out of Gauss's proofs.

When Abel was eighteen, his father died in disgrace after he had made false charges against political allies and followed it by hard drinking, which ended his political career. Since Abel was the eldest boy, the responsibility to help provide for the family fell on his shoulders. Still, he found time for mathematics, and eventually achieved a breakthrough for a problem that had occupied many mathematicians in history and would soon obsess Galois. After wrongly thinking he could solve the quintic (fifth-order) equation, he was able to prove that the quintic equation cannot be solved by an explicit algebraic expression involving its coefficients, as can be done with the quadratic equation. In 1799, the Italian mathematician Paolo Ruffini (1765-1822) had actually derived a proof of the unsolvability of the quintic equation that was overlooked for many years, so today this mathematical result is called the Abel-Ruffini theorem.

Abel came to Paris in 1826 hoping that his work would attract the interest of members of the French Academy of Sciences; but to no avail. Cauchy—who was only interested in his own investigations—was assigned to read Abel's paper, but he either ignored or lost it. Frustrated and depressed, Abel wrote from Paris to a friend: "I have just finished an extensive treatise . . . and Mr. Cauchy scarcely deigns to glance at it." He returned home, becoming weaker by the day. He had contracted tuberculosis.

In 1829, barely twenty-seven years old, Abel died. Sadly, just two days after his death, a letter arrived offering him an academic position in Berlin, where he had visited and impressed mathematicians before going to Paris.

It was in April of the year Abel died that Galois sent his own paper to Cauchy. Galois went further than Abel in explaining why a quintic equation could not be solved in radicals (meaning in terms of its coefficients), and his ideas would pave the way for a whole new area of mathematics: group theory.

On July 2 of the same year, an event with grave consequences to the life of Galois took place. His father, Mayor Nicolas Galois, had endured ruthless, ceaseless attacks from the Royalists in his hometown. They even resorted to dirty tricks that included crafting malicious poems, falsely attributing them to him, and publishing them in the local paper. The elder Galois, no longer able to endure such malevolence, committed suicide. At his funeral, a riot erupted between the two political factions.

A few days later, the young Galois again attempted the entrance examinations to the École Polytechnique. One of the two examiners, whose names were Dinet and Lefébure de Fourcy, asked Galois to explain logarithms. Distraught by the death of his father and irritated by a question he considered trivial, Galois threw the blackboard eraser at the examiner (probably Dinet).[1] Needless to say, he failed the test, to the great disappointment of his teacher.

Louis Richard then suggested that Galois apply to the "lesser" École Normale Supérieure (which, at that time, was known as the École Préparatoire, or Preparatory School). Reluctantly, Galois followed his teacher's advice, and then submitted the results of his research to the Academy of Sciences for consideration in their competition for the Grand Prize in Mathematics. The paper was received by Fourier, who took it home to read—and died. For Galois, it was yet another stroke of bad luck. The manuscript was never recovered, but some of Galois's results were later published in the June 1830 issue of Baron de Férussac's *Bulletin des Sciences Mathématiques, Physiques et Chimiques*.

To Galois's great disappointment, because the academy had lost his paper when Fourier died, the Grand Prize was awarded jointly to Abel, who had died the previous year, and the German mathematician Carl Gustav Jacobi (1804–51). After this devastating misfortune, Galois became increasingly bitter, and at the same time arrogant. He knew that his knowledge of mathematics was so good that his teachers could not even understand the new results he was deriving.

Resigned, Galois joined the entering class at École Normale Supérieure. On July 27–29, 1830, the school was locked down to prevent students from taking part in riots on the streets of Paris against Louis-Philippe of Orleans, the post-Empire reinstated Bourbon monarch. The students of the École Polytechnique, on the other hand, had no such restriction placed on them, and they made history by battling the king's soldiers in the barricades on the streets and boulevards of Paris.

Galois was greatly agitated by this development. Having been raised a Republican, he became more ardent in his political beliefs and took a Republican leadership position within the student body. According to a member of his family, his stated goal was "to defend the rights of the masses."[2] On November 10, 1830, Galois joined the Society of Friends of the People, another political body opposed to the reinstated monarchy. Around that time, he began to publicly attack the president and the administration of the university, which had supported the king. Galois also criticized the university's instruction, which he considered mediocre at best. His attacks on the administration led to his being "indefinitely suspended" from the school.

On December 1, 1830, the journal *Annales de Gergonne* published a short abstract of Galois's work, and this became the last scientific paper published in his lifetime. Four days later, on December 5, he wrote a scathing criticism of the school, accusing the professors of "mediocrity of teaching." Since he had already been suspended from the school, this new offense caused his outright dismissal, and Galois found himself on the street, with no income. In order to try to make a living, he offered a private course in mathematics on a street corner near the Sorbonne every Thursday at 1:15 p.m. On the first day, thirty students attended, and then more came, but the enterprise soon ended as student interest waned after a period of time.

In frustration Galois joined the artillery branch of the French National Guard, a Republican-leaning body of the French Civil Service. On May 9, 1831, he attended a banquet held by this unit in the Parisian restaurant Aux Vendanges de Bourgogne. Alexandre Dumas, who was also present, later described the event. Apparently, Galois raised his wineglass unexpectedly in a toast. After gaining the guests' attention by clanging a knife against his glass, he said, "To Louis-Philippe." Later, he claimed that he had added, "should he betray," but because his toast caused an immediate uproar, nobody heard it. "The fumes of the wine had removed my reason," Galois later told the police, in his defense.

The next day, Galois was arrested at his mother's house and accused of threatening the life of the king. He was sent to the Sainte Pélagie Prison, near the famous Jardin des Plantes, by the Seine. He was tried on June 15, and at his trial, he testified as follows: "Here are the facts. I had a knife I had been using to cut my food. I raised the knife while saying, 'To Louis-Philippe, if he betrays.' But the last words were heard only by my nearest neighbors because the first part of the toast elicited whistles." Galois was acquitted.[3]

On January 17, 1831, the Academy of Sciences had asked Siméon-Denis Poisson and Sylvestre Lacroix to read a paper sent to the academy by Galois. It appears that Poisson had actually encouraged Galois to resubmit the paper that had been lost when Fourier died, but when it came to actually reading Galois's paper, neither eminent mathematician understood Galois's ingenious application of permutation groups to the understanding of the relationships among roots of a polynomial equation and ultimately to the solvability of the equation by radicals. On July 4, 1831, Poisson and Lacroix issued their final report on the work of Galois: "We have made every effort to understand this proof. The reasoning is neither sufficiently clear nor developed enough to allow us to judge its correctness." Galois was very disappointed and did not supply any further explanation of his groundbreaking work. On July 14, Bastille Day, celebrated in France in commemoration of the French Revolution, Galois was again arrested. This time he was at the head of a large group of anti-Royalist demonstrators that had congregated at the Pont Neuf on the Île de la Cité in the heart of Paris. He was carrying a loaded rifle, pistols, and a dagger. Because it was his second offense, this time Galois was convicted and sentenced to six months in prison.

In December of 1831, Galois again tried to publish his work, though he was still angry that his previous papers had all been lost or misunderstood by the academy. "Egoism reigns in the sciences," he wrote to a friend. "People should study together instead of sending sealed letters to the academy."[4] In early 1832 a cholera epidemic caused the closure of the Sainte Pélagie Prison, where Galois had been held, and the prisoners were transferred to other facilities. Galois was moved to a halfway house located near the Place d'Italie, which was then in a township called Gentilly but today is in the heart of the 13th arrondissement of Paris. This was a pleasant place, as compared to a prison. It even had a resident physician to look after the health of the convicts living there. The physician was Dr. Du Motel.

In early May 1832, Galois apparently became romantically involved with a young woman named Stephanie. (We know this because "Stephanie" and "Stephanie D." appear in the margins of a number of letters he later wrote, sometimes effaced or blotted out.) She has been tentatively identified as Stephanie-Félicie Du Motel, the daughter of the physician at the halfway house. It seems that on May 14, the affair ended, and a man claiming to be Stephanie's lover challenged Galois to a duel. Alexandre Dumas identified the man as Perscheux d'Herbinville, although this identification has been disputed. Galois may have been in the National Guard and have held a leadership position with the campus Republicans at the École Normale Supérieure, but he was still

On the night before his death in 1832, Galois wrote down many of his mathematical theorems in a letter to his friend Auguste Chevalier; in the letter's last page, above, Galois writes: "I hope some men will find it profitable to sort out this mess."

a very young, inexperienced person, and naive about the ways of the world. He had never taken part in a duel before, and probably had no clear idea about how a duel was fought. Worst of all, we know from his correspondence that apparently he felt that he could not avoid a duel, even if he was certain that he had absolutely no chance of winning it. Galois seemed to know that he would die, and yet he followed through and walked into the dreadful trap laid out for him.

On May 29, the night before the duel, Galois did not sleep. In a hotel room he wrote down some of his mathematical theories, leaving out many details ("I have no time"), and sent it in a letter to his friend Auguste Chevalier. In other letters written that night to Republican comrades, the tormented genius wrote that he knew he would not survive the confrontation: "I am dying, the victim of an infamous coquette," he wrote.

At dawn on May 30, 1832, Galois went to a deserted field to meet his adversary. We can only guess how he must have felt. Apparently he did not have a second (an assistant), as duelers usually did, and he probably was not versed in the rules of this barbaric practice—by then illegal in France for two centuries. Around noon that day, a peasant found him lying at the side of a road outside Paris, mortally wounded in the stomach. His challenger and second had left the wounded Galois to die. The peasant took Galois to the Cochin Hospital, where he died the next morning at 10:00 a.m., apparently from peritonitis caused by the wound. His last words were to his brother, who was sitting at his hospital bedside: "Don't cry, Alfred," he said. "I need all my courage to die at twenty."

Galois was buried in an unmarked grave in the Montparnasse Cemetery in Paris. In response to his untimely death, his friends organized a demonstration on June 5, which led to the massacre of the Saint Merri cloister—an event that Victor Hugo described in his tragic novel *Les Misérables*.

But what really happened to Galois? Why would someone—who may never have even fired a gun—get involved in a duel when he knew he had no chance of winning it and had no strong conviction or passion about wanting to duel? The mystery of Galois's death has dogged historians of mathematics for almost two centuries, and there have been a number of theories about what actually happened. One of them is that the Royalists saw in Galois a dangerous Republican leader and wanted to kill him. According to this theory, they ensnared him with Stephanie and then challenged him to a duel for her "lost honor," knowing that he was inexperienced in duels and would likely die. Another theory was that *he* challenged the other man to the duel because he was smitten with Stephanie. In support of this theory we have the speculations that Galois's adversary was not a Royalist, but rather one of his Republican friends. Some have surmised that the adversary was a fellow Republican prisoner kept at the halfway house who had competed with him for Stephanie's affection. Another hypothesis, more outlandish and less credible than the others, is that Galois wanted to die in the duel so that his death would become a rallying point for the Republicans. We don't know what really happened, and it is likely that the true cause of Galois's death will forever remain a mystery.

GALOIS'S BROTHER ALFRED published the tragic young mathematician's papers and letters. Thanks to the work of Galois and Abel, we know that an equation of the fifth order or higher has no solution by radicals (i.e., through a formula using arithmetic operations, including the extraction of roots, on the coefficients of the equation). Galois's deep and comprehensive analysis led to yet another important theory in algebra: Galois theory, which refers to a special study of groups and fields in the context of solving equations. Galois theory is so powerful (and beautiful) that it settled problems that had occupied the minds of mathematicians for millennia.

It took two decades from Galois's death in 1832 before mathematicians finally understood the theory, which explains why all three classical problems of antiquity—the squaring of the circle, the doubling of the cube, and the trisecting of an arbitrary angle, all using only a straightedge and a compass—are absolutely impossible to solve. Galois theory is important in many areas of mathematics, including number theory, and helped in the proof of Fermat's Last Theorem in the 1990s by Andrew Wiles. It also underlies the entire theory of particle physics through the use of Galois's idea of groups and symmetries. It is hard to overestimate the influence of the work of this tormented young genius on modern mathematics.

In 1846 the French mathematician Joseph Liouville (1809–82) edited several papers written by Galois and published them together with the letter and manuscript Galois had sent to his friend Chevalier the night before the fatal duel. Galois's work was based on results obtained by Gauss that relate equations to properties of prime numbers, and on work by Lagrange on the permutations of the roots of a polynomial equation. Galois then used the term *group* for the first time in reference to mathematical groups of permutations of the roots of polynomials. When these groups had certain desirable properties of symmetry, then the polynomial equation from which they were derived had solutions obtained as radicals through arithmetic operations on the equation's coefficients. The quintic equation does not have such properties, and thus Galois showed definitively what Abel and Ruffini had been able to show in a more restricted and incomplete way. Today the theory of groups that emerged from Galois's findings is a very important part of modern abstract algebra.

The tragic death of Galois marks the end of the prominence of France in mathematics until the end of the nineteenth century. Laplace, Fourier, and Legendre all died between 1827 and 1833, and only Cauchy would continue to write papers until his death in 1857. With the deaths of most of the best French mathematicians, the leadership in mathematics was passed to England and Germany.

ALGEBRAISTS OF BRITAIN

Interestingly, it would be the British who would take over leadership in algebra, the field of Galois's fertile investigations. George Boole (1815-64) was born to a family of tradesmen in Lincoln, England. He studied Greek and Latin early in his education, and then started reading the works of Laplace and Lagrange before embarking on a career in mathematics. In 1847 he published a book called *The Mathematical Analysis of Logic*, which introduced mathematical analysis into logic, the systematic study of forms of valid deductive argument. Seven years later *Investigation of the Laws of Thought*, which extended his theory, followed. Boole introduced what is now known as *Boolean algebra*—the algebra of logic that every computer programmer knows by heart. This is the algebra of true/false statements. For example, if x and y each represent something that can either be true (1) or false (0), then one can determine the truth value of their conjunction (xy). One only needs to grasp the basic rules of multiplication to know that, if either x or y is false, then xy must be false. Our world of computers would not be the same without Boolean algebra.

Other developers of abstract algebra include George Peacock (1791-1858), who, along with Augustus De Morgan (1806-71), attended Trinity College, Cambridge. De Morgan was born in India to British parents, and after moving to England, he tried to generalize algebraic notation to virtues and vices.[5] Important contributions to algebra were also made by Arthur Cayley (1821-95), who invented the theory of matrices, a concept that is absolutely essential not only in mathematics but also in all of applied science. In 1858 Cayley wrote a paper on the theory of linear transformations. Trying to impose order on the transformations, Cayley invented the matrix. He also studied the determinant, which we think of as naturally associated with a matrix.

Cayley's close lifelong friend, James Joseph Sylvester (1814-97), was born to a Jewish family named Joseph, but he later changed his name to Sylvester (a name associated with the papacy). Sylvester also studied matrices and devised a method, called *Sylvester's criterion*, of eliminating an unknown from two polynomial equations. Together, the two mathematicians worked on developing the theory of forms. Sylvester moved to the United States and, for many years, was one of the greatest mathematicians working at Johns Hopkins University before he returned to England.

The British mathematician William Kingdon Clifford is known for his writings on the geometrical properties of gravity. Like fellow mathematician and Oxford alumnus C. L. Dodgson, a.k.a. Lewis Carroll, Clifford also wrote children's stories.

Benjamin Peirce (1809-80) of Harvard did important work on linear associative algebras, and is famous for having said, "Mathematics is the science which draws necessary conclusions."[6] His ideas were further pursued in England by William Kingdon Clifford (1845-79), who gave us Clifford algebras, which include concepts and tools that are useful in mathematics and theoretical physics. Clifford enjoyed entertaining young children with stories, much like his fellow Oxford mathematician C. L. Dodgson (1832-98). We all know Dodgson by his pen name, Lewis Carroll, the author of *Alice in Wonderland*.

WILLIAM ROWAN HAMILTON

At Göttingen, Germany, a Norwegian mathematician named Sophus Lie (1842–99) extended Galois's work to continuous groups. In a discrete group, such as the permutation groups studied by Galois, there are discrete symmetries. For example, rotating a triangle one third of the way around leaves the triangle the same, and such transformations are modeled by a discrete group of rotations. These discrete groups aid in the study of solutions of polynomial equations. Lie derived the new, continuous groups, such as the group of all possible rotations of a circle, in order to study differential equations. Unlike discrete polynomial equations, differential equations are "continuous," in that taking the derivative is a continuous operation; hence, Lie "invented" these groups to try to perform for differential equations what Galois had done for polynomial equations. Lie groups are one of the most powerful and most important mathematical tools in modern theoretical physics.

Algebraic operations are commutative—i.e., the order of the operations is unimportant ($5 \times 3 = 3 \times 5 = 15$)—but there are other types of systems in which the order of operations *is* important. For example, we know that matrix multiplication is generally noncommutative. The study of a particular noncommutative algebra was carried out by the illustrious Irish mathematician and physicist Sir William Rowan Hamilton (1805–65).

Hamilton's father and mother were intellectually inclined, but they died young. Even before they died, however, Archibald and Sarah Hamilton sent their young son to live with his uncle, Reverend James Hamilton, so that he would receive a good education at the hands of this very knowledgeable and scholarly man. With the encouragement of his uncle, who was a gifted linguist, William learned to read Greek, Latin, and Hebrew by the age of five. By the age of ten, he knew six languages, including Persian and Hindustani.

As a young man, Hamilton became friendly with the poets William Wordsworth and Samuel Coleridge. He even wrote his own poems. However, when he matriculated at Trinity College, Dublin, he turned his attention to mathematics, studying Laplace's classic *Mécanique Céleste* and other works of the great French mathematicians who had advanced mathematical methods in physics and astronomy.

While still a student, at the age of twenty-two, Hamilton was appointed the Royal Astronomer of Ireland and director of the Dunsink Observatory, as well as professor of astronomy. In 1833 Hamilton presented to the Royal Irish Academy a paper in which he introduced an algebra of pairs of real numbers and defined multiplication thus: $(a, b)(c, d) = (ac - bd, ad + bc)$. This is actually the law of multiplication of complex numbers, which Gauss understood, but for the first time a mathematician had made the law explicit in algebraic terms. Hamilton tried hard to extend this idea further, but he couldn't do it.

The mathematician John C. Baez provides a fitting metaphor for the relations among number systems, which was central to Hamilton's work:

> There are exactly four normed division algebras: the real numbers, complex numbers, quaternions, and octonions. The real numbers are the dependable breadwinner of the family, the complete ordered field we all rely on. The complex numbers are the slightly flashier but still respectable younger brother: not ordered, but algebraically complete. The quaternions, being noncommutative, are the eccentric cousin who is shunned at important family gatherings. But the octonions are the crazy old uncle nobody lets out of the attic: they are *nonassociative*.[7]

The Dunsink Observatory in Dublin, Ireland, was established in 1785. Between 1880 and 1916, when Ireland was decreed to be in the same time zone as England, the observatory recorded Ireland's official time, then called Dublin Mean Time. There are several references to "Dunsink time" in James Joyce's *Ulysses*.

The story of Hamilton's discovery of the quaternions is amazing. Hamilton was fascinated by the idea of a relationship between the complex numbers as an algebra and as a two-dimensional geometry, where multiplication by i is rotation in the plane by 90 degrees counterclockwise. He tried to go one step further and look at three-dimensional geometry as a home for an extension of the complex numbers, studying this problem at length from 1835 until 1843. By then he had been knighted by the British king.

William Rowan Hamilton, the Irish mathematician famous for the algebraic graffito he inscribed on Dublin's Brougham Bridge, could read six languages by the time he was ten years old.

In a letter to one of his sons, describing the events of October 1843, Hamilton later wrote, "Every morning in the early part of the above-cited month, on my coming down to breakfast, your (then) little brother William Edwin, and yourself, used to ask me: 'Well, Papa, can you *multiply* triplets?' Whereto I was always obliged to reply, with a sad shake of the head: 'No, I can only *add* and subtract them.'"[8] The system Hamilton was looking for simply did not exist. What he was really looking for, algebraically, was an algebra modeled by a four-dimensional geometry, not a three-dimensional one.

On October 16, 1843, Hamilton was walking with his wife along Dublin's Royal Canal, heading for a meeting at the Royal Irish Academy. As he and his wife crossed the Brougham Bridge over the canal, he suddenly had a great "aha" moment. As he described it in the letter to his son, "That is to say, I then and there felt the galvanic circuit of thought close; and the sparks which fell from it were the fundamental equations between i, j, k; exactly such as I have used them ever since."

At that moment, Hamilton stopped dead in his tracks, right at the top of the bridge. He wanted to write it all down, but he had no pencil or paper, so he picked up a nail from the ground. In what has been called the most famous act of mathematical vandalism in history, he carved into the stone surface of the bridge the following equation: $i^2 = j^2 = k^2 = ijk = -1$. (This is an extension of the special multiplication law for complex numbers, $i^2 = -1$.) The original carving is now gone, but every year his act and mathematical achievement is commemorated by mathematicians from around the world who, on October 16, replicate Hamilton's walk, culminating at the top of the bridge, where a plaque describing his discovery now stands.

The algebraic system Hamilton discovered is called the quaternions. The (unit) quaternions form a Lie group, which exhibits key symmetries found in theoretical physics. Thus all these developments that originated from the work of a young genius named Galois are not only important in pure mathematics, but also form key elements of modern theoretical physics and other fields.

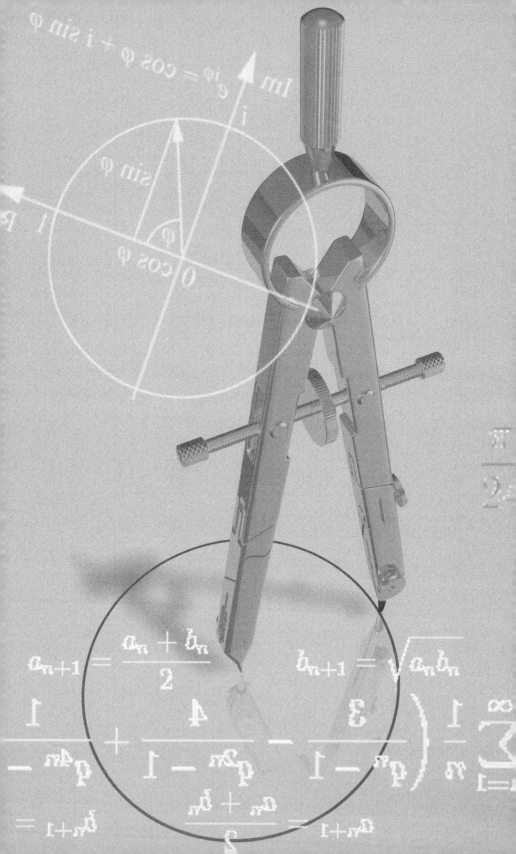

TOWARD A NEW MATHEMATICS

$$a_{n+1} = \frac{a_n + b_n}{2}$$

$$b_{n+1} = \sqrt{}$$

$$t^0 = 1$$

$$\rho^0 = \frac{\sqrt{S}}{1}$$

$$e^{i\varphi} = \cos\varphi + i\sin\varphi$$

$$\sin\varphi$$

$$\text{Re}$$

$$\sum_{n=1}^{\infty} \frac{1}{n} \left(\frac{1}{q^n - 1} - \frac{1}{q^{2n} - 1} + \right.$$

$$a_{n+1} = \frac{a_n + b_n}{2}$$

$$a_n = 1 \qquad b_0 = \frac{1}{\sqrt{2}} \qquad t_0 = \frac{1}{4}$$

———•◆•———

INFINITY AND
MENTAL ILLNESS

T he political upheaval of France in the mid-1800s had slowed the progress of mathematics there, so the center for mathematical research in the world moved to Germany. Major mathematicians worked in Berlin and Göttingen, including Richard Dedekind (1831–1916) of the Brunswick Polytechnic, who made progress in our understanding of mathematical analysis—the theoretical basis of calculus.

A key mathematician in Berlin was Karl Weierstrass (1815–97), a high-school teacher who, at age forty, published

Sunlight filters through a dense forest of spruce trees in the Harz Mountains of Germany, where mathematician Georg Cantor sought rest and relaxation. The image aptly conveys the obstacles—be they foes, doubters, or internal demons—that many mathematical geniuses face during the arduous quest toward truth.

a mathematical paper so profound that he was offered a full professorship in Berlin. There, he nurtured many students, including a young Russian woman named Sofia Kovalevskaya (1850–91). This was unusual because women were not allowed to matriculate at universities, nor did they have their own passports. She had written him from her native Russia and managed to travel to Berlin (with her parents' permission) before he agreed to take her on as a private student. Kovalevskaya received a Ph.D. from the University of Göttingen and continued to study under Weierstrass, but struggled to find employment. Eventually, however, she gained a reputation as an important mathematician, lecturing at the University of Stockholm before dying from the flu at age forty.

Another mathematician in Berlin, Leopold Kronecker (1823–91), was the spoiled son of a business magnate. He was a good mathematician—the *Kronecker delta function,* useful in many areas of mathematics, is named after him—but he believed that irrational numbers (already known to the ancient Greeks) did not really exist. He was intolerant of anyone who did not hold his views about mathematics . . . or anything else.

Into this mélange of mathematicians walked a psychologically complicated genius, Georg Cantor (1845–1918), whose pioneering work would lay the foundation for set theory and our modern understanding of infinity.

GEORG CANTOR

Arguably the title of most tormented—both internally and externally—among the greatest mathematicians in history goes to the German mathematician Georg Cantor. We know that Cantor was born in Saint Petersburg, Russia, on March 3, 1845, and that his father, Georg Woldemar Cantor, was born in Copenhagen. We also know that the father's family moved to Saint Petersburg after the 1807 British bombardment of Copenhagen, in which their home had been destroyed. The father was known to have been a devout Lutheran. Cantor's mother, Maria Böhm, was born a Roman Catholic. But otherwise, Cantor's provenance is shrouded in mystery.

Although his parents were married in a Lutheran ceremony in Saint Petersburg in 1842, there are some clues that suggest that Cantor's grandparents on both sides were Jewish. For one, Georg Woldemar's father, Jacob Cantor, married a woman whose maiden name was Meier. In addition to the fact that Cantor, Böhm, and Meier are all common Jewish names, an important piece of evidence is found in a letter Cantor wrote to a friend late in life, in which he mentioned that he had "Israelitisch" grandparents. When and how religious conversions took place, if at all, is unknown. Neither do we have much information about the grandparents, including where they were from. A tantalizing detail in the life story of Georg Cantor is the fact that, while he lived as a Lutheran, he chose to use the first letter of the Hebrew alphabet, aleph, to denote the concept he invented for infinity.

Because of the father's pulmonary illness, the family moved to Frankfurt, Germany, where they enjoyed great economic success with an international wholesale business the father founded and named Cantor & Co. But ultimately the move away from the damp and cold Baltic climate, and the financial success he enjoyed in Frankfurt, didn't help him much: some years later, Georg Woldemar would die of consumption.

Since Georg was the first of six children, his family had great aspirations for him. His uncle, Joseph Böhm, was the founder and conductor of the Vienna Conservatory. Another uncle was a noted law professor at the University of Kazan in Russia whose writings were used as the legal framework for the Russian Revolution. One of his students had been Leo Tolstoy.

Young Georg Cantor attended private schools in Frankfurt and, at age fifteen, was admitted to a gymnasium, the rigorous German high school in which mathematics and science are taught. While he was boarding at the school, in the city of Darmstadt, his father, Georg Woldemar, wrote him a letter that reveals something about how the family viewed their firstborn son:

> I close with these words: Your father, or rather your parents and all other members of the family both in Russia and in Germany and in Denmark, have their eyes on *you* as the eldest and expect you to be nothing *less* than a Theodor Schaeffer and, God willing, later perhaps a *shining star* on the horizon of science.[1]

Georg Woldemar knew Theodor Schaeffer, his son's teacher at the gymnasium, and considered him a great scholar. Georg Cantor kept this letter from his father with him throughout his life, perhaps to draw from it the strength he would need to fight against adversity from without and from within, and to surmount the immense hurdles placed in front of him in his quest to understand the nature of infinity.

The year he entered the gymnasium, Cantor confessed to his father that he loved mathematics, and the latter then used this information to push him, in letter after letter, to pursue not only mathematics, but also physics and astronomy. The father told his son about his nighttime dreams. In one of them, he was looking at the stars and marveling at their apparent infinitude. Some historians of mathematics, including E. T. Bell, have identified Georg Woldemar's extremely high expectations as a possible cause of his son's mental problems later in life.

Cantor performed well in his school examinations—better in mathematics and the sciences than in geography, history, and the humanities—and the school administration therefore recommended that he pursue a university course of study in the sciences. He was admitted to study mathematics at the renowned Swiss Polytechnic Institute in Zurich, and, after a short stint there, was able to transfer to the University of Berlin, the undisputed world leader in mathematics research at the time. Cantor took courses from many of the great mathematicians in Berlin, but the subject he chose as his concentration was the theory of numbers, and in 1867 he wrote a brilliant dissertation in this area. Weierstrass influenced

Cantor taught at the University of Halle, Germany, from 1869 until the end of his life. The school was founded in 1694; this lithograph depicts one of its "new" buildings under construction in 1836.

him greatly, leading him to the study of infinity. Upon obtaining his doctorate, he took the first position he could get—that of a *privatdozent* (private tutor) at the University of Halle.

Cantor's life ambition was to become a professor at Berlin, but his views on infinity and irrational numbers quickly came into conflict with those of Kronecker, who was a very influential member of the mathematics faculty in Berlin. The latter continually stood in his way to a professorship at this prestigious university, and Cantor, despite his genius, would remain condemned to teaching at the second-tier University of Halle for the rest of his days.

Cantor was undeterred by attacks against his mathematics, however. Weierstrass, the hulking bigger-than-life presence at Berlin, pioneered approaches to numbers that included quantities that were infinitesimally small and used them to lay a foundation for mathematical analysis. And the German mathematician Dedekind worked in the same direction, defining irrational numbers using the idea of a cut (*schnitt* in German), which was essentially a sequence of rational numbers that converged to an irrational one. In the work of these two men, Cantor found solace from the mounting attacks of Kronecker, a diminutive man with a bad temper who resembled a small barking dog, ever harassing Cantor for his work on infinity.

Cantor started with the work of Weierstrass on the convergence of number sequences. Weierstrass,

Karl Weierstrass, above, Cantor's doctoral-thesis adviser at the University of Berlin, also served as adviser to Arthur Schoenflies, Cantor's first biographer.

like Newton, Leibniz, and the ancient Greeks two millennia before them, used the concept of potential infinity—the idea commonly used by any student of calculus today. Saying "let h approach zero," or "let n go to infinity," are limit arguments that use potential infinity without ever actually exhibiting an infinite quantity. But at a certain point in his work, Cantor decided to address the question of the actual existence of infinite quantities. It was here, in fact, that Cantor had taken a giant step forward, beyond anything that had passed before him. And it was this act of blasphemy, as it were, that engendered the ire of his detractors—chief among them Kronecker, but also Kronecker's Berlin colleague Ernst Eduard Kummer.

From work on the convergence of series as some quantity "goes to infinity," Cantor made the leap to actual infinity—a concept only glimpsed by few of his predecessors, among them Galileo and Bernhard Bolzano (1781–1848). While on house arrest by order of the Inquisition, Galileo turned his attention to pure mathematics and noted that he could draw a one-to-one correspondence between all the integers and all the squared numbers simply by squaring a number, as we've seen. Hence, he could "count" all the integers against all the squared integers and find that these two sets were of the same size. But it would seem that there are more integers than squared integers. This is the paradox of infinity: an infinite set can be put into a one-to-one correspondence with a proper subset of itself. (Recall the "Infinite Hotel," which demonstrates this idea.) Bolzano did the same with continuous numbers. He used the function $y = 2x$ to "count" numbers on an interval and show that an interval twice as large as another could still be viewed as having the same "number of points" as the original interval.

In Halle, Cantor settled down to the mediocre life of an academic in one of Germany's lesser institutions of learning and research. While Berlin was teeming with ideas and great mathematicians with whom to discuss them, at Halle there was a dearth of intellectual stimulation. Nevertheless, he found a comfortable life there. In 1875, Cantor married Vally Guttmann, his sister's friend, who came from a Jewish Berlin family. They began to raise their own family and, with the help of his father's money, purchased a spacious house with large windows that let in much light. But mathematics is best pursued in a social, intellectual environment where ideas can be shared, discussed, learned, and refined.

Cantor, who appears in this undated portrait, suffered his first mental breakdown in 1884.

Because he was in a provincial town and was virtually the only gifted mathematician in his department, Cantor had to work in a vacuum—save for the powerful ideas he had brought with him from the courses he had taken from Weierstrass and others in Berlin.

Weierstrass had further developed a mathematical result using ideas on infinity and irrational numbers pioneered by Bernhard Bolzano, which we know as the Bolzano-Weierstrass theorem. This powerful theorem states that an infinite sequence in a closed and bounded space attains a limit point inside the space itself. Thus, if you define an irrational number as a limit of rational numbers converging to it, then the number is as well defined as any other number, since it is in the original space (consisting of numbers) in which the sequence was defined. The theorem builds on ancient Greek ideas about numbers and is also rooted in Richard Dedekind's definition of irrational numbers using cuts of rational numbers.

THE CONTINUUM HYPOTHESIS

Cantor came to the idea of *actual* infinity not by looking at numbers but rather by considering sets. Cantor started from Weierstrass's idea of defining irrational numbers as limits of sequences of rational numbers. He then decided to look at sets of irrational numbers in intervals of numbers, and then looked for the set of all limit points of any given set of limit points in an interval. Each such set is a "derived set," as it is derived from the original set of limit points. Thus he created a sequence of sets of limit points. He then asked himself, "When does the derived set of limit points become empty?" That is, he wanted to know if there ever was a case where the set of limit points derived in this way—as limit points of limit points—eventually becomes exhausted.

Cantor was fascinated by the process of deriving infinite sets of numbers from other sets of numbers. Since he used sets in his analysis, he became known not only as the person who explained infinity, but also as the founder of the modern theory of sets. Using the idea of counting, in which we establish a one-to-one correspondence between two sets of numbers—e.g., to count four sheep, we establish an unambiguous one-to-one correspondence that associates the number one with the first sheep, the number two with the second, and so on—Cantor invented an ingenious tool that allowed him to "count" all the rational numbers, even though that set is infinite.

The diagonal array below demonstrates Cantor's "diagonalization proof" that the infinite number of all the rational numbers is exactly the same as that of the seemingly smaller set of all integers.

Note that the arrows weave through the infinite array, demonstrating how all the integers and the fractions can be listed. The arrows, extending from number to number as shown, produce a one-to-one corre-

```
1/1   1/2 ⟩ 1/3    1/4 ⟩ 1/5    1/6 ⟩ 1/7    1/8 ⟩ ...
 ↓ ↗   ↙   ↗   ↙   ↗   ↙   ↗
2/1   2/2   2/3    2/4   2/5   2/6   2/7   2/8    ...
      ↙   ↗   ↙   ↗   ↙   ↗
3/1   3/2   3/3    3/4   3/5   3/6   3/7   3/8    ...
 ↓ ↗   ↙   ↗   ↙   ↗
4/1   4/2   4/3    4/4   4/5   4/6   4/7   4/8    ...
      ↙   ↗   ↙   ↗
5/1   5/2   5/3    5/4   5/5   5/6   5/7   5/8    ...
 ↓ ↗   ↙   ↗
6/1   6/2   6/3    6/4   6/5   6/6   6/7   6/8    ...
      ↙   ↗
7/1   7/2   7/3    7/4   7/5   7/6   7/7   7/8    ...
 ↓ ↗
8/1   8/2   8/3    8/4   8/5   8/6   8/7   8/8    ...
  :     :     :     :     :     :     :     :
```

spondence between all possible fractions of whole numbers (including integers, such as $2/1$ or $4/2$, and other ways of representing them) and all the integers by pairing them, an integer and a fraction, all the way to infinity. What we do is order all the fractions (i.e., rational numbers), and *count* them: 1, 2, 3, . . . They comprise an infinite set, but they are counted one-to-one against the (infinite) totality of the integers. (The various representations of the same number, such as 2, $4/2$, $8/4$, etc., do not alter the argument.) So although there are infinitely many rational numbers and infinitely many integers, the number of numbers in each set is identical!

So are all infinite sets equal to each other in their number of elements? In 1874 Cantor proved that the answer to this question is *no*—there are infinite sets that are "more infinite," or larger, than the set of integers and rational numbers. Cantor used another incredibly ingenious technique to prove this result: the age-old Greek idea of proof by contradiction. Cantor asked himself a simple question: Can I list all the *real* numbers as I did with the rational numbers? Lists can be infinite, as long as you can provide a rule that accounts for the progression from one number to the next. In the case of the rational numbers, Cantor had been able to come up with a rule for the list without actually listing every number. Attempting to list all the real numbers—meaning rational and irrational numbers, all together—Cantor began by simply trying to list all the real numbers between zero and one. If such a list existed, then, without any particular order, it would look something like this:

$$0.1\ 3\ 4\ 2\ 8\ 8\ 7\ 6\ 3\ 1\ 5\ \ldots$$
$$0.3\ 6\ 2\ 6\ 6\ 5\ 3\ 9\ 8\ 9\ 1\ \ldots$$
$$0.7\ 8\ 5\ 5\ 1\ 1\ 2\ 3\ 4\ 9\ 2\ \ldots$$
$$0.2\ 2\ 5\ 3\ 7\ 7\ 9\ 1\ 2\ 1\ 4\ \ldots$$
$$\vdots$$

We assume that such a list is exhaustive—i.e., that it includes all the real numbers between zero and one. If this were the case, we would be able to associate each of the infinitely many real numbers with a unique integer, and thus "count" them (in principle; although in practice this would take "forever"). If we can prove that the real numbers between zero and one are indeed "countable," as mathematicians say, then the argument could perhaps be extended to the set of all real numbers. Its *order of infinity* would then be the same as that of the integers and rational numbers.

But in fact, this is not the case. Cantor was able to show that, whatever the numbers on this list may be, we can *always* find a new number that is not on the list! For example, if we isolate the number that makes up the *diagonal* of all the randomly ordered numbers on the list above, meaning we take the first decimal from the first number on the list, the second decimal from the second number, the third from the third number, and so on, we get: 0.1653 . . . Now add *one* to each of these decimals (with the understanding that a nine changes to a zero). We now come up with a brand-new number: 0.2764 . . . This new number could not *possibly* be on the infinite list of numbers above because it differs from every number on the list by at least one decimal. Thus real numbers are *uncountable,* and their order of infinity is higher than the order of infinity of the integers and rational numbers.

Cantor gave a name to the "infinity" of the integers and the rational numbers: *aleph-zero*. However, he did not know how to classify the order of infinity of the real numbers. It was clearly higher than the order of infinity of the integers and rational numbers, but was there a level of infinity in between these two? Cantor hypothesized that the answer was no, and thought that the order of infinity of the real numbers was indeed the next order of infinity. He called the next-highest order of infinity–the one following that of the integers and rational numbers–*aleph-one*. Then he spent the rest of his life vainly trying to prove this statement–namely, that the continuum of real numbers is of the next level of infinity after that of the rational numbers. This statement has become known as the continuum hypothesis.

———————————◆———•——————————————

THE RELATIONSHIP BETWEEN Georg Cantor and Leopold Kronecker was one of the most complex and troubling in the history of science. Kronecker came from a wealthy Jewish family of bankers and businessmen, and he never had to worry about his income. Cantor had been Kronecker's student at the University of Berlin, and at first Kronecker was supportive of his former student and enthusiastic about his first steps in mathematical research in Halle. However, as Cantor began to explore the deep recesses of infinity and discover highly counterintuitive, bizarre properties of infinite sets and numbers, Kronecker became irritated and went on the attack.

In the work of Georg Cantor, we find the bizarre contradictions that lie in the depths of the foundations of mathematics. These contradictions would ultimately hamper Cantor's attempts to obtain a complete grasp on infinity. We will look at one familiar paradox, Russell's paradox, which shows that there is no set containing all sets. Put simply, there is no universal "basket," as it were, containing all baskets (sets) inside it.

Suppose that such a universal set does exist. Now consider the set of all sets that do not contain themselves. Does this set contain itself? If it does, then it doesn't, and if it doesn't, then it does. Thus what we have is a paradox and it implies that a universal set cannot exist. We can illustrate such impossibility in an easier way with the well-known Barber of Seville paradox. The Barber of Seville shaves all men in Seville who do not shave themselves. Does the barber shave himself? If he does (as the barber), then he doesn't (as a man of Seville), and if he doesn't (as the barber), then he does (as a man of Seville). Similar paradoxes to this simplified one plague the very foundations of mathematics.

We know that Kronecker had a philosophical distaste for infinity, irrational numbers, and the continuum, but was his virulence toward Cantor perhaps attributable in part to the fact that he was upset that a former student was making his way in mathematics in spite of him? We don't know. Either way, Cantor felt the wrath of Kronecker and his Berlin cronies: "It is as if the Berlin mathematics department turns into darkest Africa, with lions roaring and hyenas screaming," he once wrote a friend to describe the reaction to his name being mentioned at the department.

Weierstrass understood his genius, and Dedekind became his friend, but the disproportionately influential Kronecker continuously stood in his way, taunting Cantor with declarations such as: "God made the integers, all the rest is the work of man." Irrational numbers (such as π or the square root of 2) are explained using infinite series, and Cantor was a master of these. Today every graduate student takes for granted that there are various levels of infinity—as Cantor has taught us—but to Kronecker this was sheer heresy.

In the economically depressed city of Halle—later, part of East Germany—Cantor was a respected professor. But every time he published a paper, there was a ruckus in Berlin. He took it badly. Between his lack of progress on the continuum hypothesis and the constant attacks by Kronecker and other "lions and hyenas," Cantor's mental health deteriorated.

Cantor seemed to view infinity in an almost religious sense. He deeply believed that infinity was the realm

Cantor's bitter rival and former teacher Leopold Kronecker, left, called Cantor a scientific charlatan and a corrupter of youth.

of God. Infinity included the transfinite numbers—aleph-zero, aleph-one, and any others—and, beyond the transfinite numbers, there was an unreachable level of infinity that he named the Absolute. According to Cantor, the Absolute was God himself. As Cantor continued to explore the idea of infinity and the levels he had found infinity to have, his conflict with Kronecker intensified and became even more personal.

The bitter rivals did make one attempt at reconciliation, however. Cantor liked to vacation in Germany's Harz Mountains, a small mountain chain west of Halle whose highest peaks do not exceed three thousand feet. The villages, forests, and streams of the Harz Mountains offered a restful environment in which to relax and discuss mathematics. Indeed, Cantor had met a number of mathematicians in these mountains, and during one of his stays, he gathered his courage and wrote a letter to his former professor offering a meeting in one of the mountain villages. To his surprise Kronecker accepted the invitation. The two men then met and spent a few days discussing mathematics, and it seemed that they might have reached some kind of understanding. But soon after they departed—Cantor to Halle and Kronecker to Berlin—the confrontation between them flared up again.

On June 29, 1877, Cantor wrote a letter to Dedekind in which he exclaimed that he had discovered something so bizarre that he found it too shocking to comprehend—even if he had proved it on paper. "I see it," Cantor wrote, "but I don't believe it!" What Cantor discovered was that a central concept in mathematics and everyday life—dimension—is completely irrelevant when it comes to the powerful concept of infinity.

Cantor drew a square and a line segment on a Cartesian coordinate system and asked himself: "Are there more points on the square than on the line segment?" The intuitive answer is "of course," but Cantor analyzed the situation using his customary machinery of searching for a one-to-one correspondence between the points on a square and the points on a line segment. To his great surprise, he succeeded in drawing such a correspondence.

Cantor chose for his line segment the interval of numbers between zero and one, and for each side of his square also an interval of real numbers between zero and one. Now, each point on the square is given by its two Cartesian coordinates, and since each coordinate is between zero and one, a point is given by the pair $(0.a_1a_2a_3a_4a_5a_6 \ldots, 0.b_1b_2b_3b_4b_5b_6 \ldots)$.

Cantor defined the transformation from the square to the line segment as *alternating* the decimal expansion of each of the two coordinates. Thus a number on the line segment would be of the form $0.a_1b_1a_2b_2a_3b_3 \ldots$ This establishes a one-to-one correspondence between every number on the square and every number on the line segment. Here, the line segment between zero and one is a proper subset of the square with side being the zero-to-one interval, which proves that dimension doesn't matter when it comes to the level of infinity of a set of numbers: a two-dimensional object (the square), has the same *cardinality* (i.e., the same order of infinity) as the one-dimensional object, the line segment.

Cantor tried to get his paper on the irrelevance of dimension published in *Crelle's Journal*, but Kronecker caught wind of the submission and actively sought to prevent its publication. After hearing only silence from the journal's editor for several months, Cantor became suspicious. When he inquired, he found out that, indeed, his enemy was behind the stonewalling. Kronecker had told the editor that Cantor's paper dealt with empty concepts that did not really exist and that the mathematical public had to be protected from meaningless concepts like infinity and irrational numbers. *Crelle's Journal* eventually did publish Cantor's paper a year later, but the mathematician decided to send his future papers to other publications that he hoped might be immune to the influence of his foe.

Cantor befriended a Swedish mathematician named Gosta Mittag-Leffler, who edited the mathematics journal *Acta Mathematica*, and this journal then became the home of Cantor's later papers. Although Cantor's work was warmly appreciated by an editor who understood his work, Kronecker was not ready to give up his relentless assaults. After seeing Cantor's papers in *Acta Mathematica*, Kronecker tried to befriend Mittag-Leffler himself by pretending that he had a paper he wanted to publish in the journal. As it turned out, there was no such paper, but Cantor became very angry with Mittag-Leffler once he suspected that Kronecker was successful in turning his friend and supporter against him. This reaction almost lost him a key ally.

Stress was taking its toll on Cantor's health, and in May 1884 he had the first of his many nervous breakdowns. The attack lasted two months, during which he was completely unable to work. At that time, Cantor had been working on what we now know is an impossible problem: the continuum hypothesis.

CANTOR'S ELDEST DAUGHTER, Else, was nine years old at the time of his first nervous breakdown. She and other members of the family were so shocked by the sudden change in her father's behavior that she remembered it vividly years later, as she recounted to Cantor's first biographer, Arthur Schoenflies. At the onset of the problem, Cantor became very agitated and could not communicate with people. He stayed in bed for weeks, during which time he neglected mathematics altogether and read Shakespeare. Eventually, he became convinced that he had come up with a new finding: that Shakespeare's plays were actually written by Francis Bacon. It later came out that he had come upon this idea through a book he had found at an antiquarian bookshop in Leipzig, which described Bacon as a great poet rather than a scientist, and Cantor had assumed that the book he had found was unknown. Psychologist Nathalie Charraud hypothesizes in her book *Infini et Inconscient: Essai sur Georg Cantor* (Infinity and the Unconscious: An Essay on Georg Cantor) that Cantor may have seen himself as a character in a Shakespearean tragedy. In a haze of mental illness, rage, and hurt from the onslaught against him from Berlin, Cantor was living in an unreal world. He decided to leave mathematics altogether and asked the university's administration to allow him to transfer to the philosophy department, but his request was turned down.

After publishing two pamphlets at his own expense, both of which argued that Bacon was the author of Shakespeare's plays, Cantor returned to mathematics and to his unsuccessful attempts to prove the continuum hypothesis. In 1899, after a concentrated effort at proving this impossible hypothesis, Cantor suffered another mental breakdown and was taken to the Halle Nervenklinik, a mental care facility in the city, for treatment.

Today we recognize that Cantor probably suffered from bipolar disorder, in which periods of depression alternate with periods of elation. During the "high" periods he would work as if in a frenzy, and during the "low" periods he would be completely immobilized. Today there are effective medicines for treating this problem, but in Cantor's time treatment consisted of making the patient take long, hot baths. Cantor had a pleasant private room in the facility, with high windows that let the sun in—he was a respected professor at the University of Halle and was treated as such. He often walked the wooded grounds of the hospital, enjoying nature. And he rested. After a few months Cantor was released, but he did not feel up to resuming his work and wrote lengthy letters to the Ministry of Education asking to be relieved of his duties as professor and given a post as librarian—at the same salary, he insisted. Apparently, the ministry ignored these requests.

There ensued a cycle of intense periods of work on the continuum hypothesis and frequent hospitalizations. Unfortunately, Cantor was at the Nervenklinik at the time of the 1900 Congress of Mathematicians, held in Paris, where the renowned German mathematician David Hilbert presented a set of ten problems—later expanded to twenty-three—that he considered the most important in mathematics and hoped would be solved in the coming century. The first problem on the list was Cantor's continuum hypothesis.

Cantor was well enough in 1904 to attend the Third International Congress of Mathematicians, held in Heidelberg, Germany. He came accompanied by his two daughters, Else and Anna-Marie. His youngest son, Rudolph—a gifted musician—had died at age thirteen after years of ill health, and Cantor had also lost his mother not long before the congress. At the congress the Hungarian mathematician Jules C. Koenig presented a paper claiming that the second order of infinity was not any of Cantor's alephs. Sitting in the audience, Cantor became enraged and started ranting against the speaker, causing a great commotion in the audience before his daughters could calm him down.

Nevertheless, Cantor was such an astute mathematician that—once he was again able to think clearly—he saw immediately that Koenig had made improper use of one of the lemmas (preliminary mathematical results) in his work. Shortly afterward the German mathematician Ernst Zermelo, who, along with Cantor and Abraham Fraenkel, would become recognized as one of the founders of modern set theory, proved that Koenig's work was indeed flawed because of his misuse of that lemma.

Over the next decade and a half, Cantor went through cycles of hospitalization and recuperation, and in June 1917 he was admitted to the clinic for the last time. On January 6, 1918, his emaciated body was found in his room in the Nervenklinik. Apparently, food shortages resulting from World War I had affected the hospital's supplies, and Cantor died of starvation.

Cantor's pioneering work opened the door to our modern understanding of infinity, but his goal of solving the continuum problem could not be achieved—not by him and not by anyone else. In 1937 the Austrian logician Kurt Gödel proved a result that, once completed by Paul Cohen of Stanford in 1963, established that Cantor's continuum hypothesis cannot be proved or disproved within our system of mathematics (called Zermelo-Fraenkel set theory). Incidentally, Gödel, too, suffered from mental problems throughout his life: he died from self-starvation.

FOURTEEN

UNLIKELY HEROES

y the turn of the twentieth century, humanity had made tremendous leaps in industry and communication, but prejudice was still an ever-present reality. Whereas earlier mathematicians struggled amid religious and political upheaval or against internal demons, two mathematicians born in the late nineteenth century fought the constraints of poverty and sexism in their quests to be heard as mathematicians. They came from unusual backgrounds and lived atypical lives in a world where, despite a proliferation of egalitarian principles and social ideals, the sciences were still dominated by wealthy white men. These trailblazers were the Indian genius S. R. Ramanujan and the brilliant German mathematician Emmy Noether.

London's Burlington House is depicted in this engraving as it appeared in 1854, when it became the headquarters for the Royal Society of London for Improving Natural Knowledge, founded in 1660. The building serves as a symbol, perhaps, of the intellectual establishment to which iconoclastic mathematicians like Srinivasa Ramanujan—who, in fact, was elected a Fellow of the Royal Society in 1918—sought to belong.

RAMANUJAN

The mathematician Srinivasa Ramanujan (1887–1920) was born in the village of Erode, south of Madras in Tamil Nadu, southern India, to a poor family. His father worked in a small store, and his mother sang at a temple. A younger brother died from a childhood illness at just three months of age, and two other siblings also died as infants. Ramanujan himself contracted smallpox when he was two but recovered from it. At age five he enrolled in school in Kumbakonam, a town nearer to Madras, where his family lived. He did not enjoy school.

At a young age Ramanujan exhausted the mathematical knowledge of his teachers at school and independently read books on mathematics. Although he had no formal training in mathematics, he had such an amazing aptitude for it that by age twelve he had worked out new solutions to problems in number theory and analysis. Astonishingly, he seemed to be able to come up with mathematical facts and ideas in a complete intellectual vacuum. India had a long tradition, going back to the early Middle Ages, of producing important mathematical results without proof. And like some other Indian mathematicians, Ramanujan cared little about formal proof. He simply derived beautiful mathematics as if out of thin air—most of them identities and equations. In 1902 he learned the method the Italian mathematicians had found in the sixteenth century for solving cubic equations, and he derived on his own a method of solution for quartic equations. He was not aware of the impossibility of solving quintic equations by radicals, so he spent time, in vain, trying to derive a formula for it.

At his high-school graduation in 1904, Ramanujan was awarded the Rao prize for outstanding achievement in mathematics, having obtained grades that were higher than the maximum possible at the school. At the Government College in Kumbakonam, where

Ramanujan, whose likeness appears on this 1962 commemorative stamp, said in an interview shortly before his death, "As a child, I was considered slow-minded, as my verbal abilities did not come into play until I was three years old."

he studied on scholarship, Ramanujan also performed amazingly well in mathematics but showed no aptitude for anything else. He therefore lost the scholarship and moved to another town by himself, later going to Pachaiyappa College in Madras to study. Because of his poor performance in other areas of study and some health problems, he failed to graduate. Nevertheless, his independent study of mathematics produced many results.

But leaving college without a degree made Ramanujan depressed. Through an Indian tradition allowing such unions, he married a ten-year-old girl, Janaki Amal, with the stipulation that the marriage would not be consummated until she came of age. They did not live together, and Ramanujan continued to live in abject poverty. He began looking for some kind of job to support himself, hoping to find employment as a clerk. Ramanujan finally obtained the needed references for employment when he showed his mathematical work to several mathematics professors at local universities. His work was so astonishingly novel that some of these professors at first doubted that it was his own. Once he showed them how he derived his equations, however, they understood that he was not a fraud, and their enthusiastic letters of reference even enabled him to receive some financial support to work on mathematics.

Ramanujan published in the *Journal of the Indian Mathematical Society* a problem that he challenged other mathematicians to solve. His riddle was to find the answer to the infinite sum of these nested square roots:

$$\sqrt{1 + 2\sqrt{1 + 3\sqrt{1 + \cdots}}}$$

Six months passed, and no one had come forward offering a solution, so Ramanujan revealed the answer: 3.

In 1912 Ramanujan finally obtained a position as a clerk in the Madras Port Trust. He performed his job so efficiently that he had time left over to do more research in mathematics and publish papers. Seeing how brilliant he was, Ramanujan's friends and associates showed his work to English mathematicians to try to gain their support for the struggling young man. Unfortunately, these attempts were completely unsuccessful.

In January 1913 Ramanujan wrote a letter to the prominent British mathematician G. H. Hardy (1877–1947) at Cambridge University, including nine pages of his original mathematical work. Hardy looked at the paper and initially thought it was fraudulent; he thought someone must have copied the work of some mathematician from a journal without citation. He recognized some of the results as mathematical derivations that had been obtained by others and known in the West. Others made no sense to him. But he was intrigued. When he read the pages again, he realized that one result he didn't understand was obtained from work on hypergeometric series, previously studied by Euler and Gauss. He was so impressed and stunned by the theorems that he later said, "They defeated me completely; I had never seen anything in the least like them before."[1] These theorems had to be true, he concluded, because "if they weren't, nobody would have had the imagination to invent them."

Hardy showed the papers to his colleagues, and they were equally stunned. Then he wrote back to Ramanujan, expressing interest in his work and asking for proofs of some of the theorems. Ramanujan was elated to receive a response and wrote to Hardy, "I have found a friend in you, who views my labors sympathetically." Eventually, Hardy invited him to Cambridge University. When the invitation arrived, the local education board decided to give Ramanujan a grant to work at the University of Madras, in hopes of keeping him in India. His parents apparently objected to Ramanujan's proposed move to England, and he sadly turned Hardy down. Hardy was disappointed, and their relationship cooled somewhat, but Hardy tried again. This time around, Ramanujan was ready—his mother had had a dream in which the family deity told her to allow her son to leave.[2]

On March 17, 1914, Ramanujan boarded the *Nevasa* at Madras, arriving in London almost exactly a month later. He moved to an address very close to Hardy's rooms at Cambridge, and the two men met daily to go over Ramanujan's amazing theorems. Hardy had already received in letters from Ramanujan more than one hundred theorems, and

British mathematician G. H. Hardy, seen in this ca. 1927 photograph, was an early supporter of Ramanujan's, and remained his friend and advocate until the younger man's premature death at the age of thirty-two.

Ramanujan had brought many more with him. Looking at the theorems, Hardy could see that some were already known and others were false, but many of them were new breakthroughs.

In later years Hardy would say that his greatest achievement in mathematics was discovering Ramanujan. He considered him a mathematician of Euler's caliber. But Hardy was also an excellent mathematician himself. In his famous book, *A Mathematician's Apology*, he wrote about the life of a pure mathematician. Hardy had never been interested in applications, concerning himself instead with bringing rigor and abstract beauty to British mathematics. Despite his tremendous ability, he was generally shy and reserved. For some reason, he hated to look at himself in a mirror. It was reported that when he traveled, he would cover the mirrors in the hotel rooms where he stayed. Hardy never married, but he maintained several close relationships with people in his life. His friendship with Ramanujan was prime among them.

Ramanujan was awarded a doctorate for his work at Cambridge, and he eventually became a Fellow of Trinity College, a Fellow of the Cambridge Philosophical Society, and in 1918 was elected Fellow of the Royal Society. But his health suffered. He had endured a variety of ailments throughout his life, and in England he felt worse. After five years in Cambridge, he required frequent hospitalizations. Although he was believed to have been suffering from tuberculosis, later findings suggested a parasitic infection affecting his liver.

Ramanujan was said to be a friend of every integer, and no incident demonstrates his love for numbers better than an exchange he had with Hardy not long before his death. While he was lying ill, Hardy came to visit him. In an attempt to lift the ailing mathematician's spirits, Hardy led off with a comment about a number, numbers being Ramanujan's favorite topic: "I came here in a taxi with a very boring number: 1729." To his surprise Ramanujan gathered whatever strength he still possessed, jumped up in bed, and cried, "No, Hardy, no, Hardy—it's a very interesting number! It's the smallest number expressible as sums of two cubes in two different ways!" (The number $1,729 = 10^3 + 9^3 = 1^3 + 12^3$.) From this event, the mathematical study of *taxicab numbers*—the smallest numbers that can be expressed as sums of two (positive) cubes in n distinct ways—emerged. To date, only the first six taxicab numbers have been found.

Shortly after Hardy's visit, Ramanujan's condition worsened. In despair, he decided to return home to India, thinking that the warm weather there would help him improve. In February 1919 he left for India, arriving in Madras the following month. Unfortunately, he was already too sick to recover. On April 26, 1920, in Kumbakonam, Ramanujan died. He was only thirty-two years old. His widow, Amal, continued to live in Madras until her death seventy-four years later, in 1994.

FELIX KLEIN AND THE ERLANGEN PROGRAM

One of the most prominent mathematicians in Germany in the late 1800s was Felix Klein (1849–1925), who taught at the University of Erlangen, in Bavaria. After attending the gymnasium in his home-town of Düsseldorf, he matriculated from the University of Bonn, where he had plans to earn a degree in physics.

Under the direction of Julius Plücker, the chair of mathematics and experimental physics, Klein studied many types of geometry that were emerging at the time. He also studied the connections between geometry and group theory—the area discovered by Galois. In 1868, he received his doctorate. When Plücker died in the middle of writing a book on geometry, Klein took up where his former pro-

fessor had left off and completed it. The resulting work, *New Geometry*, included ideas on such emergent geometries as the Riemannian geometry of the brilliant German mathematician Georg Friedrich Bernhard Riemann (1826–66), which is very important in both mathematics and theoretical physics, as well as the

At the University of Göttingen, Germany, Felix Klein supervised the first Ph.D. thesis ever written by a woman in the mathematics department. The degree was awarded to Grace Chisholm Young in 1895.

non-Euclidean geometries developed by Gauss, János Bolyai (1802–60), and Nikolai Lobachevsky (1792–1856). Klein visited Paris and was greatly impressed by the methods of group theory developed at that time. Groups can be connected with geometries. Once a group associated with a geometrical object can be identified, this gives the mathematician information about the geometry of an object that might otherwise not be discoverable.

While lecturing at Göttingen, Klein met Sophus Lie, with whom he collaborated in research on groups and their properties. When he became a professor of mathematics in 1872, at the age of twenty-three, Klein gave an address that has become well known around the world. In this lecture, he inaugurated the *Erlangen Program,* through which he hoped to classify geometries by their associated groups of symmetries. Three years after he became professor at Erlangen, Klein married Anna Hegel, the granddaughter of the famous German philosopher Georg Wilhelm Friedrich Hegel.

In 1882 Klein discovered a nonorientable surface now called the Klein bottle. It is an extension to three dimensions of the previously known Möbius strip, a nonorientable surface named after the German mathematician August Ferdinand Möbius, who discovered it in 1859. Both surfaces are shown below.

The two-dimensional Möbius strip (left), and the three-dimensional shape known as the Klein bottle (right), are nonorientable surfaces.

An ant walking on the top face of a Möbius strip would eventually find itself walking on the bottom face of the same strip, and then again on the top face, and so on as it continued. The Klein bottle, by comparison, is a tube looped back through itself to join its other opening, so it has no well-defined "inside" or "outside." (It doesn't really exist in Euclidean space.)

The Georg-August-University of Göttingen was founded in 1737 by Elector George Augustus II of Hanover—a.k.a. King George II of Great Britain. This late-nineteenth-century lithograph shows the school's Auditorium Maximum, built in 1826.

In 1886 Klein moved to the more renowned University of Göttingen, where leading research in mathematics was taking place. But the shameful story of how Weierstrass in Berlin had to resort to teaching the brilliant Sofia Kovalevskaya on the side because women were not permitted in classes weighed on enlightened German academics at Göttingen. Klein, in particular, was pained by the German higher education system's treatment of women.

In 1893 he managed to convince the University of Göttingen to admit women, which was a great step forward. Under the newly passed law, Klein had the opportunity to teach a young English-woman named Grace Chisholm Young (1868–1944), who had been a student of the inventor of the matrix, Arthur Cayley, at Cambridge. At Göttingen, Young earned her Ph.D. working under Klein. Sexism was still rampant in Europe, however, and her first works were published under her husband's name.

In 1895, through the influence of Klein, Göttingen hired David Hilbert (1862–1943), who, along with Klein, made Göttingen even more prominent in mathematical research. As mentioned previously, David Hilbert gave the keynote address at the 1900 Congress of Mathematicians, in which he listed the ten problems he hoped mathematicians would solve in the twentieth century.

Klein and Hilbert were both interested in group theory and mathematical physics. Then, in 1900, Max Planck (1858–1947), a prominent German physicist working at the Berlin Academy, discovered the quantum principle. The new discovery virtually shouted for help from mathematicians and increased interest in mathematical physics. Coincidentally, the mathematical "space" that best describes the quantum world happens to be *Hilbert space*, a concept developed by David Hilbert within the context of pure mathematics. This was the milieu of pure and applied mathematics in Germany when the brilliant Emmy Noether appeared on the scene.

EMMY NOETHER

We now turn to the life of one of the most interesting and important female mathematicians in history. Writing in *The New York Times* after her death, Albert Einstein described her as the most important woman in mathematical history. She lived and worked in Germany during the same period that Ramanujan worked in England and India, but whereas Ramanujan's contributions were in analysis and number theory, hers were in abstract algebra and in the application of algebra to theoretical physics. The two celebrated Noether's theorems laid a strong mathematical foundation for conservation laws in physics, the first theorem being essential in quantum field theory and the second in general relativity.

Emmy Noether (1882–1935) was born to a Jewish family in Erlangen. Her father, Max Noether, came from a wealthy merchant family and taught himself mathematics. As professor of mathematics at the University of Erlangen, he was involved in the emerging field of algebraic

The mathematician Emmy Noether, whose theorems deal with the conservation of energy in physics, appears in this undated photograph.

geometry, which would later be vastly expanded and extended by Alexander Grothendieck, who is discussed in the next chapter.

Emmy grew up in a happy home in Bavaria; she loved to dance and enjoyed music. At the public girls' school she attended, she performed very well but showed no interest in the area that would become her life's work. She excelled in language, studying English and French, and prepared for a career as a language teacher in German state schools. She even took and passed the teacher's qualifying examination.

But somewhere along the way, the young woman changed her mind and decided not to teach. Instead, she received special permission—as was required of all women who attempted to study at a German university—to sit in on courses. She audited mathematics courses at the University of Erlangen, and then moved to Göttingen and audited mathematics courses taught by some of the most famous mathematicians, including Hilbert and Klein. She also sat in on courses given by the chair of the mathematics department, Hermann Minkowski (1846–1909), a Russian-born German mathematician who had previously been a professor at the Swiss Federal Institute of Technology in Zurich (ETH). There, in the 1890s, Albert Einstein had been a student of his, and Minkowski—who had done key work on number theory, studying sums of squares and the geometry of quadratic forms—acquired a strong interest in mathematical physics. Minkowski is the mathematician to whom we owe the geometrical understanding of Einstein's four-dimensional space-time.

Hilbert, who was developing his own interest in mathematical physics, had lured Minkowski away from Zurich to the mathematics department at Göttingen just a year before Emmy Noether arrived at the university in 1903. When she returned to Erlangen, the university gave her a degree based on all the courses she had taken unofficially at Erlangen, Göttingen, and Nuremberg, where she studied for a while. She then started her doctoral work at Erlangen and was awarded a Ph.D. in 1907, after presenting her dissertation on invariants. (Roughly speaking, invariants are things that remain mathematically the same after some change is made to the parameters of the problem.)

Having obtained her doctorate, Noether was well qualified for a position at a university, but the persistent sexist atmosphere in Germany prevented the brilliant young woman from being able to even apply for a job. This depressed her greatly, but she began helping her father—by then in poor health—with his research. She also started publishing papers on her own, which were so well received that she was invited to join a number of European mathematical societies, including the German Mathematical Association, which had been founded by Georg Cantor.

EINSTEIN'S REVOLUTION

In 1905, the world of science was shocked when a twenty-six-year-old patent clerk in Bern, Switzerland, named Albert Einstein showed that space and time are interlinked as something called *space-time*. Likewise, Klein, Hilbert, and Minkowski were shocked by the news that time slows down for a fast-moving object and that the speed of light is the only "invariant" in Einstein's theory, the *special theory of relativity*. In an effort to explain Einstein's revolutionary new findings within mathematics rather than within the language of theoretical physics, Minkowski—Einstein's former professor from Zurich—wrote his paper on four-dimensional space-time.

The three great mathematicians at Göttingen also realized that Einstein's new physics was "invariant" under the action of a group. Thus Galois and Lie both entered the story of relativity. The Lorenz group, named after the Dutch physicist Hendrik Lorentz (1853-1928), was identified as the group of all transformations that leave the Minkowski four-dimensional space-time invariant, in that it allows us to correctly change the coordinate system to reconcile what different observers see. A larger group of transformations comprise the Poincaré group, named after the French mathematician Henri Poincaré (1854-1912). Incidentally, Poincaré has been called "the last universalist" because he was a master of so many different areas of mathematics.[3] (Interestingly, a celebrated conjecture Poincaré had made in 1904 about the geometry of three-dimensional surfaces was proved almost a century later by the reclusive Russian mathematician Grigori Perelman, who then refused both the prestigious Fields Medal and the $1 million Millennium Prize for his breakthrough.)

The need for new mathematical tools in physics became more acute ten years after special relativity was proposed. In 1915

Einstein shocked the world again by presenting his *general* theory of relativity, which included gravity and thus changed our understanding of Newton's work, taking us far beyond its realm to extremes of speed and gravitational force in which Newtonian mechanics does not work. Ever since finishing his special theory of relativity in 1905, Einstein had been working hard to try to apply the relativity principle to Newton's theory of gravity. In 1907 he tried an approach that didn't work and then realized that he needed new mathematics. His friend Marcel Grossmann lent Einstein his notebooks from classes they had both taken at the ETH in Zurich (Einstein, apparently, was bad at taking notes). Through them Einstein was led to non-Euclidean geometry, the work of Riemann, and finally to obscure results by Italian mathematician Gregorio Ricci-Curbastro (1853–1925), who had studied with Felix Klein, and his collaborator Tullio Levi-Civita (1873–1941). The two had developed the "absolute differential calculus"—a calculus method that used tensors, which are generalizations of the matrices invented by Cayley.

In Switzerland and Prague, Einstein labored on his tensors, trying to make them yield a mathematical system that would explain Newtonian mechanics in a relativistic context. Then he was invited to work at the Berlin Academy and moved back to his native Germany. By this time, he had realized that gravitation curves space-time, and in 1914 he sent the Berlin astronomer Erwin Freundlich to the Crimea in order to prove the validity of his emerging theory of relativity by making observations of stars during a total solar eclipse. Einstein had predicted that starlight passing by the sun would curve around it, distorting the star's position. An eclipse in the Crimea offered the opportunity to prove such curvature. But World War I intervened, and Freundlich was arrested by the Russians. By the time of his release, the eclipse had already occurred.

In 1915 Einstein accepted an invitation to visit Göttingen and agreed to give a talk to the mathematics department about his ongoing research on a tensor equation, in which the elements are arrays of variables, for general relativity. David Hilbert was in the audience, and he took copious notes on what Einstein was writing on the board. Then he went to his office and did perhaps the most unkind thing he had ever done: starting from Einstein's work, he tried to derive the right equation. When he thought he had it, he sent his equation of general relativity, which used tensors of dimension 14 (meaning fourteen rows of variables), to a journal for publication.

At the same time, Einstein made the breakthrough he had been looking for and sent *his* general relativity equation, with ten-dimensional tensors, to a journal. Research by Jürgen Renn of the Max Planck Institute in Berlin has established the exact time line of both paper submissions in November 1915 and the fact that Hilbert indeed tried to beat Einstein to the finish line. But the mathematician got his equation wrong, while the physicist got it right!

This photograph of Einstein was taken during a lecture he gave in Vienna in 1921.

As it turned out, Hilbert's cumbersome fourteen-dimensional equation lacked an essential property: invariance. Relativity requires a certain kind of invariance called *general covariance*, which means that the physics an equation describes should not depend on the frame of reference from which it is observed. Whether you are looking at a rocket launch from the north or from the south, you should see the same physical process. Einstein's equation was also far more elegant and compact than Hilbert's. The Pythagoreans, who worshipped the number 10—their *tetractys*—would have been proud of it.

Somehow, the mathematician who had proven several theorems about invariance failed to imbue his equation with this key property. Hilbert clearly needed someone who understood invariance even better than he did if he wanted to understand the mathematics of Einstein's general theory of relativity. It was then that he and Klein remembered the exceptional young woman who had sat in on their courses a decade earlier and who was now publishing papers without an academic home.

BY 1915 NOETHER WAS a famous mathematician in her own right, and her papers were read with interest throughout the world. They dealt primarily with algebra. Like Hardy, who invited Ramanujan to come and work with him in Cambridge, Klein and Hilbert invited Emmy Noether to work with them at Göttingen.

Noether arrived at Göttingen and began her work on invariance in mathematical physics. Meanwhile, Klein rallied to get her appointed a professor at Göttingen, but he had to struggle with the administration until 1919, when his request was finally granted. While she was at Göttingen, and during visits to Moscow State University in Russia in the 1920s, Emmy Noether accomplished incredible achievements in mathematics. In 1915, she presented the two Noether's theorems, which show a paramount connection between the concept of symmetry in mathematics and the very important conservation laws in physics. More specifically, the theorems show that when the Lagrangian (the mathematical expression invented by Lagrange that captures the elements of a physical situation) enjoys a certain kind of symmetry—i.e., when it is invariant under the action of a Lie group, such as the group of all rotations of a circle—then the physical system modeled by the Lagrangian implies a conservation law. Thus Noether's theorems explained the conservation of energy, momentum, and electric charge in the language of group theory.

But Noether's work went far beyond mathematical physics. She made important contributions to Galois theory, to many other areas of abstract algebra, and to topology. Noether was, in fact, the greatest algebraist of her time. She worked prodigiously, having little social life other than close contact with her students. In her lectures she talked very fast, a reflection of her very quick thought processes. She also suggested many research topics to her students, and many of her solutions to key problems in abstract algebra appeared in print as the work of her students, even though she had played a role in deriving the results.

Unfortunately, when Hitler came to power in 1933, Noether was summarily dismissed from her position so that the university could comply with the newly passed race law forbidding Jews from holding academic jobs in Germany. After failing to obtain a position at Moscow State University in the Soviet Union that year, she accepted a visiting professorship at Bryn Mawr College in Pennsylvania and came to the United States. Two years later she died from an infection following an operation to remove a tumor. Meanwhile, Emmy's brother Fritz Noether—also a mathematics professor dismissed by the Nazi regime—obtained a position at the University of Tomsk in Siberia, but his tenure didn't last. In 1937 the Soviet government falsely accused him of being a German spy and sent him to a Soviet camp, where he was executed in 1941.

Emmy Noether expanded the role of women in the mathematical community. She advanced the field of algebra, and used algebraic methods to address key problems in physics. Utilizing Lagrangian methods and Euler's discovery of the calculus of variation, she tied together the work of the algebraists with calculus and topology to form a framework for explaining the workings of the physical world. Likewise, Ramanujan's prolific, unconventional results inspired a great deal of research in the twentieth century, finding application in the work of André Weil, who, at the time of Noether's death, was engineering one of the biggest pranks in mathematical history.

FIFTEEN

THE STRANGEST
WILDERNESS

fter Germany's defeat in the First World War, the world center of mathematics returned to France, and Paris became home to more import-ant working mathematicians than any other city. These "new" mathematicians, politically disillusioned by the Great War (and certainly constituting part of Gertrude Stein's *génération perdue*, or Lost Generation), were highly cynical and angry. They hated the "establishment," both political and educational. And they disparaged the older mathematicians, whom they viewed—perhaps unfairly at times—as morally corrupt. This new sociopolitical milieu in France made fertile ground for the greatest revolution in mathematics—and the invention of a person who never existed.

The Chapelle de St. Eutrope, or Chapel of St. Eutropius, in Castanet-le-Haut, France, is situated in the Languedoc-Roussillon region, not far from the Pyrenees, where some suspect that the mathematician Alexander Grothendieck lives in hiding. The village of Andabre can be seen in the valley at right.

NICOLAS BOURBAKI

There was once, in French history, a general of Greek origin named Charles Bourbaki, who in 1871 led French forces to one of the most humiliating defeats in the war against the Prussians, and later attempted suicide—unsuccessfully. His story captured the imagination of Raoul Husson, a history buff who also happened to be a third-year math student at the École Normale Supérieure (ENS), which, since its acceptance of Évariste Galois a century earlier, had become a very prestigious university in Paris. Every year, the third-year students at ENS would play a trick on the entering freshmen in mathematics, and in 1923, it was Husson's turn to plan and carry out the annual prank.

Husson, wearing a long fake beard and pretending to be a professor, walked into a room full of freshman and wrote on the board: "Theorem of Bourbaki. Prove the following . . ." The "theorem" was completely nonsensical, and the poor souls sat there for an hour scratching their heads and worrying that they were failing their very first assignment in a mathematics course at this august university.

The pseudonym Nicolas Bourbaki originated with student pranks carried out during the early twentieth century at Paris's École Normale Supérieure, the courtyard of which is seen here.

At the same time, other students played tricks on passersby. Below the university area, south of the Luxembourg Gardens and toward the Paris Observatory, passes one of the major boulevards of Paris, the Boulevard du Montparnasse. Long before the boulevard was built, this area took its name from "Mont Parnasse," the facetious nickname that French students in the seventeenth century gave to the large garbage heap that festered there. ("Mont Parnasse" is the French version of Mount Parnassus, the sacred peak near Delphi in central Greece, where Apollo was worshipped in ancient times.)

On a sidewalk of the Boulevard du Montparnasse, a student stood on a podium and lectured to the crowd about the poor conditions in the (fictitious) nation of Poldevia. Other students would tell passersby that they were collecting money to aid the people of Poldevia, who were so poor that they couldn't even afford trousers. As they said this, a student appeared from behind the podium and was introduced as the prime minister of Poldevia. He was wearing only underwear.

André Weil (1906–98) loved these two student tricks. He came from a wealthy Jewish family in Alsace who had chosen French citizenship when the residents of Alsace-Lorraine were given a choice between becoming citizens of France or Germany. His father was a surgeon who had worked for the military, and the family owned a beautiful apartment on the Left Bank in Paris, right next to the beautiful Luxembourg Gardens. It was there that André and his sister, Simone—who would become a famous Catholic philosopher—grew up.

Weil was extremely smart, but perhaps not a genius on the level of Cantor or Galois. A precocious child, he found that school was too easy—and besides, he was more interested in having a good time. He was especially keen on playing tricks on unsuspecting people. Immediately after earning his doctorate in mathematics from the École Normale Supérieure, the young man was sent to India to become the chair of the department of mathematics of one of the state universities—a position customarily given to a senior member of the faculty, not a brash young Ph.D. transplanted from Europe. Weil was still in his early twenties, and the accolades went to his head.

One day in 1930, Weil suggested to his Indian friend Damodar D. Kosambi, who had just obtained his doctorate in mathematics from

Harvard and returned to his home country, that he write a completely nonsensical paper and see if he could get it accepted by a professional mathematics journal. He also told him the stories about Bourbaki and Poldevia from his student days in Paris. Kosambi wrote some mathematical nonsense and titled the paper "On a Generalization of the Second Theorem of Bourbaki," the inside joke being that the first theorem of Bourbaki had been written on the blackboard at ENS in Paris some years earlier, supposedly to be proved by the hapless entering students. In the introduction to the paper, Kosambi wrote that the theorem was attributed to "the little-known Russian mathematician D. Bourbaki, a member of the Academy of Sciences of Poldevia, who was poisoned during the Russian Revolution." He and Weil had a good laugh, and then he sent it to the *Bulletin of the Academy of Sciences of the Provinces of Agra and Oudh Allahabad* for review. To their surprise the journal accepted the paper!

In this mood of irreverence, trickery, and contempt for prevailing academic systems, Weil returned to France some years later and, in 1933, took a teaching position at the University of Strasbourg. There he reunited with Henri Cartan, whom he had known during his student days in Paris. Cartan, now a fellow faculty member, kept complaining to Weil that he hated the textbooks they were expected to use in their classes, which were chosen by the Ministry of Education. Weil, with Cartan and with four other disgruntled young mathematicians from various French universities, decided to meet at a trendy café at the corner of the Boulevard Saint-Michel and rue Soufflot called A. Capoulade. Weil's autobiography, *The Apprenticeship of a Mathematician* (1992), describes how this happened:

Several members of the Bourbaki group and their friends posed for this photograph in 1938. From left to right, they are: Simone Weil, Charles Pisot, André Weil, Jean Dieudonné, Claude Chabauty, Charles Ehresmann, and Jean Delsarte. Not present are Henri Cartan, René de Possel, and Claude Chevalley.

One winter day toward the end of 1934, I came upon a great idea that would put an end to these ceaseless interrogations by my comrade [Cartan]. "We are five or six friends," I told him some time later, "who are in charge of the same mathematics curriculum at various universities. Let us all come together and regulate these matters once and for all. And after this I shall be delivered of these questions." I was unaware of the fact that Bourbaki was born at that instant.[1]

Meeting at the café, and in spirit continuing Weil's long string of pranks, the six friends—Henri Cartan, Jean Dieudonné, René de Possel, Claude Chevalley, Jean Delsarte, and André Weil—founded a secret group. They assumed the collective pseudonym Nicolas Bourbaki, and even faked a birth certificate for this individual. They also gave him a godfather: the prominent French mathematician Jacques Hadamard (1865–1963). They also pretended that Nicolas Bourbaki had a daughter, Betti, whose marriage they announced with printed wedding invitations. A nonexistent person thus became "real."

HENRI POINCARÉ VERSUS JACQUES HADAMARD

Henri Poincaré was undoubtedly one of the greatest mathematicians of his time, and as noted earlier, he made important contributions to many parts of mathematics. These include the invention of algebraic topology, discussed in his book *Analysis Situs* (1895). The celebrated Poincaré conjecture, mentioned earlier and proved by Grigori Perelman in 2003, is a statement in topology. It says that every closed three-dimensional manifold that is homotopy-equivalent to a three-dimensional sphere is, indeed, a sphere. The three-dimensional sphere is a generalization to one more spatial dimension of the usual (two-dimensional) surface of a ball (which is what we call a sphere). In simple terms, what the Poincaré conjecture says is that such a *three-dimensional* sphere is the only kind of bounded (i.e., not extending to infinity) three-dimensional object that has no holes in it (i.e., it is continuous, and has no tears or any kind of discontinuity in it).

Poincaré introduced the *fundamental group* in topology—an algebraic device for studying topological properties. His work was thus somewhat related to that of Felix Klein at Göttingen, who studied groups in geometry. The two men were, in fact, on friendly terms until Klein objected to Poincaré's high opinion of the work of the Prussian mathematician Lazarus Fuchs. Poincaré worked on functions he called Fuchsian, since Fuchs had done some related work, but today we call these mathematical objects *automorphic functions*. Poincaré made a stunning connection between these functions and non-Euclidean geometry. The story Poincaré told in his book *Science and Method* (1908) about making his discovery is one of the most interesting tales about how some minds make mathematical discovery:

> At that moment I left Caen, where I then lived, to take part in a geological expedition organized by the École des Mines. The circumstances of the journey made me forget my mathematical work. Arrived at Coutances, we boarded an omnibus for I don't know what journey. At the moment when I put my foot on the step, the idea came to me, without anything in my previous thoughts having prepared me for it: that the transformations I had made use of to define the Fuchsian functions were identical with those of non-Euclidean geometry. I did not verify this, I did not have time for it, since scarcely had I sat down in the bus than I resumed the conversation already begun, but was entirely certain at once. On returning to Caen I verified the result at leisure to salve my conscience.

Poincaré worked in complex function theory, differential equations, probability, and mathematical physics. He also founded chaos theory when he discovered chaotic dynamics while working on the celebrated three-body problem. Poincaré seemed to favor intuition over rigorous mathematical derivations and proofs, and he wrote very popular books about science and mathematics.

Thus, although he was a truly great mathematician, the post-Great War generation of young French mathematicians—including

This photograph of the mathematician Henri Poincaré served as the frontispiece to the 1913 edition of his book *Last Thoughts*.

the founders of the Bourbaki group—disapproved of his approach, viewing it as inexact and unrigorous. The Bourbaki mathematicians wanted as their role model a mathematician who stressed proofs and rigor; hence, they chose as their "godfather" a prominent French mathematician other than Poincaré (who had, in fact, died in 1912 at the age of fifty-eight). Not surprisingly, André Weil played the deciding role in this choice.

André Weil's dissertation adviser at the École Normale Supérieure in Paris was the mathematician Jacques Hadamard (1865–1963). Hadamard was born in Versailles to a family of Jewish descent and received his doctorate from the École Normale Supérieure, where he later became a professor. In addition to Weil, Hadamard also directed the doctoral work of three other French mathematicians who would become well known in the field: Paul Lévy (one of the "fathers" of modern probability theory), Maurice Fréchet (who did key work in analysis, topology, and abstract spaces), and Szolem Mandelbrojt (who later became a member of the Bourbaki group and whose nephew, Benoit Mandelbrot, discovered fractals).

Hadamard's work spanned many areas, including complex function theory, analysis, number theory, probability, functional analysis, the calculus of variations, differential equations, and differential geometry. The eponymous *Hadamard inequality* and Hadamard matrices attest to his important work in linear algebra. In number theory Hadamard proved a major theorem called the *prime number theorem,* which, as we recall, was proposed by Gauss and says that the number of prime numbers less than x tends to infinity at the same rate as the function $x/\ln x$. This theorem was independently proved by another French mathematician, Charles de la Valée Poussin. The proof of this theorem had been attempted but not completed by Bernhard Riemann. In related work, Hadamard won the Grand Prix des Sciences Mathématiques for work associated with the very famous, still-open problem in mathematics called the Riemann hypothesis (mentioned earlier), which is a statement about the zeros of Riemann's *zeta function* arising in number theory within the context of the complex plane.

In the 1890s a young French military officer of Jewish descent was falsely accused of treason and sentenced to life imprisonment on Devil's Island in French Guiana. Hadamard, whose wife was a relative of the accused, Alfred Dreyfus, became personally involved in what would become known as the Dreyfus affair. He wrote in defense of Dreyfus—an act of courage that united him with Henri Poincaré, who had done the same—and became a leading proponent of the young officer's innocence. Thanks in part to his efforts, the false charges were ultimately dismissed and Dreyfus was named an officer in the Legion of Honor in 1906. The celebrated French author Émile Zola was not as fortunate. After his criticism of the sham Dreyfus trial appeared in the press in the form of his now-famous article, *J'Accuse*, he fled France for Britain to avoid a jail sentence (which was later commuted).

Hadamard succeeded Poincaré as the chair of mathematics at the French Academy of Sciences and spent much time organizing and editing the mathematical work left behind by Poincaré after the mathematician's untimely death in 1912. Four years later, at Verdun, during the height of World War I, Hadamard lost both his older sons to the war (a third son would die in World War II).

There is a little-known story involving Weil, Hadamard, and the famous French anthropologist Claude Lévi-Strauss. With the Nazi conquest of Paris in 1940, these three Jewish men found it extremely dangerous to stay in France (Weil had also deserted the French army), so they traveled to New York. Hadamard had a visiting professorship at Columbia University, and Lévi-Strauss was trying to make sense of the complicated marriage laws of the aboriginal Murngin tribe of northern Australia. At some point he came to the realization that the problem was highly mathematical, so he visited Hadamard at Columbia and asked him for help. Hadamard listened to him sympathetically, and then replied, "Mathematics has four operations: addition, subtraction, multiplication, and division—marriage is not one of them." Lévi-Strauss was disappointed by Hadamard's response, but he didn't give up. Some days later he found Weil, Hadamard's former student. Weil studied the Murngin marriage rules the anthropologist showed him, and he found that, indeed, the problem was deeply mathematical and very complicated. (According to Murngin marriage laws, a man must marry one kind of cousin, if she exists, but is absolutely forbidden to marry a woman who happens to be another kind of cousin. Similar rules hold for women. This leads to the existence of sets of people within the tribe who, in turn, are either must-marry or taboo.) Weil was intrigued, and he ended up solving the problem (of determining the long-term structure of a society that follows these intricate marriage laws) using the abstract mathematics of group theory. It was an applied piece of work he remained very proud of throughout his life, even though he was otherwise a pure mathematician.[2]

The six young men of Bourbaki—among the best mathematicians of France at that time—continued pursuing practical jokes, and would also use them to promote their goals. But they also had a serious common purpose: to overthrow the stagnant educational regime of the time. They wanted mathematical education to be completely revamped and all the old textbooks thrown away. And beyond this, they had a loftier and far more ambitious goal: to *redo* all of mathematics.

Nicolas Bourbaki began publishing mathematical papers and textbooks, which were intended to replace the old and ineffectual ones. These were all high-quality publications, and the group, which grew over time and in later years included the noted French mathematicians Jean-Pierre Serre and Pierre Cartier, held regular meetings in French resort towns. Bourbaki became so convinced of "his" existence that he once wrote a letter to the American Mathematical Association requesting membership. But the secretary of the AMA at the time, Ralph P. Boas, was no fool. He wrote back saying, "I understand that this is not an application for membership from an individual," adding that if Mr. Bourbaki wanted to become a member of the American Mathematical Association, he would have to reapply as an association (and pay the much higher membership fee!).[3]

Bourbaki's many achievements include laying the foundation for the "new math" approach instituted in the United States at midcentury. In a sense Bourbaki was able to redo mathematics, as "he" had intended, restructuring the basics of the discipline. Its foundation was set theory—the discipline initiated by Georg Cantor a few decades earlier.

The series of books published by Nicolas Bourbaki was called *Elements of Mathematics*, recalling Euclid's *Elements*—the first, ancient Greek foundation of mathematics. The Bourbaki books recast much of mathematics in a new structural sense that demands logic and strict proof of all results. In helping to make mathematics more rigorous, more precise, and more proof-based, Bourbaki essentially tore down an edifice and built it back up from scratch. Of course, this was the view of Bourbaki and its adherents. Other mathematicians downplay its importance.

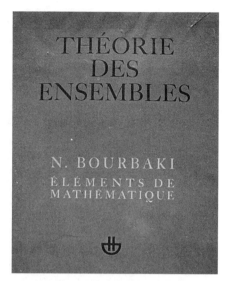

Bourbaki's *Théorie des Ensembles* (*Theory of Sets*), volume 1 of their six-volume series *Elements of Mathematics*, was first published in France in 1938. This photograph shows the cover of the first edition. Note that Bourbaki uses the singular word *"mathématique"*—not *"mathématiques"*—to emphasize the unity of mathematics.

After one of the Bourbaki summer meetings in 1939, André Weil became worried about the prospects of war in Europe. Technically an officer in the French army since his graduation from the university, Weil would have been expected to remain in France to await his call-up. But in an effort to avoid military service, he abruptly left for Finland with his wife, Éveline, and spent some time by a lake not far from the Russian border. People became suspicious, and eventually a police inquiry was made. The police found in his possession documents bearing the name of Nicolas Bourbaki, including invitations for Betti Bourbaki's wedding. They assumed he was a spy. Éveline returned home, and Weil's wartime misadventures began, including an arrest, a deportation to France, imprisonment, and conscription as private (after he was stripped of his officer rank for leaving). He then deserted from the front and escaped to the United States. Thereafter, Bourbaki's leader would remain based in America, at the University of Chicago and later at Princeton, but he continued to attend meetings in France.[4]

The Bourbaki organization still exists, but its influence is now diminished. There is still a Bourbaki Seminar in Paris that meets regularly, and technically there is still a membership in the organization, which includes about forty people. But by the 1960s, the Bourbaki group had done much of what it had set out to do—it had revolutionized the mathematics curriculum and published its *Elements*. It was now time for new ideas in mathematics.[5]

ALEXANDER GROTHENDIECK

While the Bourbaki group was most active, Weil's leadership would often be challenged by an unusual individual: the immensely brilliant mathematician Alexander Grothendieck (b. 1928), a one-time member of the secret group but more often a lone genius who changed much of modern mathematics and redid the theory of algebraic geometry.

What we know about Grothendieck, who lives in an unknown location, comes mostly from his ruminative autobiography, *Recoltes et Semailles* (Harvesting and Sowing), which includes writings on mathematics and other topics. The book has not been published, but was circulated in manuscript form among a number of his friends. In early 2010 Grothendieck sent a letter from his hiding place to a friend in Paris, demanding that all his writings be removed from circulation. Consequently, a Web site called the Grothendieck Circle, maintained by some of the mathematician's admirers, pulled all electronic copies of the manuscript out of cyberspace, along with other writings by Grothendieck. This move was only the latest in the bizarre story of the most mysterious mathematician of our time.

Very few other writings on the life of Grothendieck exist, but the most important is a paper by a member of Bourbaki who knew him well. Therefore, much of the information on Grothendieck in this chapter is drawn from the article "A Mad Day's Work: From Grothendieck to Connes and Kontsevich—The Evolution of Concepts of Space and Symmetry" by Pierre Cartier, which is based almost exclusively on what Grothendieck had recounted to Cartier.[6] Other sources include articles by mathematics professor Winfried Scharlau[7] and American Mathematical Society editor Allyn Jackson.[8]

Alexander Grothendieck's father, Alexander "Sacha" Shapiro, was born on October 11, 1889, in a town near the meeting point of the borders of Russia, Belarus, and Ukraine. It was a region within the larger "pale of settlement" of Imperial Russia, where Jews were allowed to live. This smaller Russian-speaking borderland has changed hands over time, and in it a large Jewish community flourished until the Second World War, when most of its members were exterminated.

Cartier claims, based on Grothendieck's recollections, that all Shapiros (regardless of variations in the spelling of the name) came from this limited geographical region, and that Shapiro's father, Grothendieck's paternal grandfather, belonged to the Hasidic community there. At some point, his son Sacha shed the restrictions of Orthodox Judaism and, politically, began to embrace revolutionary ideas.

Shapiro took part in the ill-fated revolt against the tsar in 1905, when he was only sixteen years old. He spent a dozen years in jail in Siberia and was released in 1917—just in time for the Russian Revolution against the same tsar, Nicholas II. In February of that year, the so-called Menshevik Revolution erupted in Saint Petersburg, and in October the Bolshevik Revolution began. After much bloodshed, the Russian monarchy was abolished, and Shapiro assumed a leadership position in the Socialist-Revolutionary Party of the Left. At first his party was allied with the Bolsheviks, but it soon went its own way and became opposed to Lenin, who

Mathematician Alexander Grothendieck's father, Sacha Shapiro, took part in the Russian Revolution of 1917, when street protests like this one—in which a Bolshevik regiment marches through the streets of St. Petersburg—were a familiar sight. Shapiro was a well-known leader of a leftist political party at the time.

Like these female soldiers, Alexander Grothendieck's mother, Hanka, fought with the Republican forces against Francisco Franco in the Spanish Civil War.

then took his revenge on its members. During this turbulent period, Shapiro's actions as a revolutionary leader earned him a place in John Reed's famous book about the Russian Revolution, *Ten Days That Shook the World*.

According to Cartier, Shapiro went on to take part in the Béla Kun revolution in Hungary, followed by uprisings in Berlin and Munich. He also joined anarcho-communist guerrilla leader Nestor Makhno's Black Army in Ukraine against the Red Bolsheviks and the White Tsarists. He seemed to go wherever revolutions were erupting, but Russian authorities were bent on capturing him. He was arrested once again by the Russian police, and this time condemned to death. Just before his execution, he managed to escape from prison but as a result lost an arm. As a wanted person, he had no choice but to flee Russia in the middle of the night. He did so with the help of a woman named Lia, who, with him, slipped across the border into Poland. Sacha Shapiro, a one-armed person on the run, changed his name to Alexander Taranoff. For the rest of his life, he would be a stateless person.

Alexander and Lia crossed borders as if they didn't exist, hiding in France, Belgium, and Germany. Shapiro/Taranoff became an ardent anarchist and had a string of relationships with women. While still a Russian revolutionary, he had secretly married a Jewish woman named Rachel and fathered a son, Dodek. In Western Europe he had other relationships while living with Lia. At some point the couple moved to Paris, where they stayed for two years.

Through his anarchist connections, Sacha met many people who were involved in the social upheavals of the time. He left Lia and traveled to Germany, where he became deeply involved with the local anarchists and met a young fellow activist named Hanka Grothendieck. Born in

Hamburg in 1900, she wrote for a local newspaper about the German sex trade, interviewing girls and women who were being exploited. Sacha and Hanka moved in together, even though she was then married to a journalist named Alf Raddatz and had a daughter by him. (Her husband was continually traveling in connection with his anarchist activities, and the couple had separated.) On March 28, 1928, Sacha and Hanka had a son they named Alexander. Since they were, in fact, married to other people, and since Sacha was living under an assumed name, the baby was given his mother's married name, Raddatz, which was changed to Grothendieck after her divorce.

Sacha and Hanka's time together did not last long. When Alexander was only five years old, his father had to leave. Hitler had come to power in Germany, and an anarchist Jew like Sacha was doubly threatened. Thus Sacha escaped to Paris, where he worked as a street photographer. For a one-armed man, this was a difficult job, and he barely made a living. After some time, Hanka placed her son in foster care at the home of a Lutheran pastor named Wilhelm Heydorn, who was fervently anti-Nazi. Her daughter was placed in an institution, and she joined Sacha in Paris.

In 1936—the year of the Spanish Civil War—no anarchist in Europe could resist the urge to join the Spanish Republicans in their war against Generalissimo Francisco Franco. Sacha and Hanka were among thousands who went to fight. When Franco won, thousands of Republican volunteers again infiltrated the border across the Pyrenees back into France. Many of them were caught as they crossed and sent to internment camps in the foothills of the mountains, but Hanka and Sacha managed to avoid capture. Sacha found his way back to Paris, while Hanka fled to the southern French city of Nîmes, where she found temporary employment as a teacher until she rejoined Sacha in Paris. All through this time, Alexander received no visits from his mother's many relatives in Germany, and no news about his parents.

In 1939 the Heydorns decided it was too dangerous for them to continue keeping a boy they thought "looked Jewish," so just as World War II began, eleven-year-old Alexander was sent alone by train to Paris, where he was reunited with both his parents. But again, alas, their time together did not last long. French authorities were bent on arresting anarchists and revolutionaries, whom they viewed as a danger to the security of the country, and eventually the family was caught. Sacha was then sent to the notoriously brutal camp Vernet, near the Pyrenees, and from there to Auschwitz, where he died in 1942. Hanka Grothendieck and her son were moved from camp to camp. Shortly after World War II was over, she died from tuberculosis contracted at the camps.

ALEXANDER GROTHENDIECK, a child living with his mother in hunger and privation in wartime concentration camps, taught himself mathematics. Children in the camps were often provided with some education, but it was sporadic and of generally low quality. It is therefore extremely surprising that a boy who grew up in such dire circumstances rediscovered by himself the entire mathematical theory of measure.

When he visited Paris in 1947, Alexander was told by the professional mathematicians he met at the university that the theory he had developed so amazingly well all by himself had already been discovered in 1905 by the French mathematician Henri Lebesgue. It was a major accomplishment by an untrained young genius, but it was only the first of many.

At war's end, Grothendieck was finally able to pursue his education at the university and catch up on the mathematics others had

learned in school. Possessing a vision in mathematics that is unparalleled, he quickly caught up and surpassed everyone else. At advanced seminars in Paris later on, the young maverick would often ask provocative questions that older and far more experienced mathematicians could not answer. It became clear to many excellent mathematicians in Paris—the center of the world of mathematics—that Grothendieck was a rising leader in the field.

Grothendieck developed a strong interest in topological groups. These are the groups of abstract algebra that

This photograph of the mysterious and elusive mathematician Alexander Grothendieck was taken in Montreal in 1970.

have a topological structure, meaning that they admit notions of continuity, as do lines, planes, and surfaces. In order to learn about them, he attended a lecture by Charles Ehresmann, a member of the Bourbaki group. Ehresmann was a recognized world expert on topological groups, but the young and naive Grothendieck did not know that. Right after the talk, he rushed to the speaker and asked, "Are you an expert on topological groups?" Ehresmann, a modest man, replied, "Well, I know something about topological groups." Grothendieck, turning to walk away, said, "But I need a *real* expert in this field!"[9]

Grothendieck received his formal education in mathematics at the University of Montpellier in southern France. As Cartier describes it, he was an outstanding student at the university: "It has often been said that he was ahead of his teachers and that he was already exhibiting a taste for extreme generalization in mathematics."[10] In 1948, having obtained his degree, Grothendieck arrived in Paris with a letter of recommendation from one of his professors in his hand; it was addressed to the great French geometer Élie Cartan. By then Cartan was quite old and in poor health, so Grothendieck was taken under the care of Élie's son, Henri Cartan, a member of the Bourbaki group. Eventually, Grothendieck's doctoral adviser would be Laurent Schwartz, another Bourbaki member, who, coincidentally, had visited the camp at which Grothendieck was interned during the war and gave lectures to the few students there. Under Schwartz, Grothendieck wrote a brilliant dissertation on topological tensor products for a doctorate from the University of Nancy. It was published in 1955.

Grothendieck went on to develop many new concepts in mathematics, including the idea of a sheaf and many other topics in algebraic geometry. He also changed the way we view space, as Cartier explains:

Our definition of points in space goes back directly to the ancient Greeks. Euclid defined a point as something that has no extent at all. In modern particle physics, for example, an elementary particle such as the electron is a "point particle," with no extent or internal structure, although it has mass. In the seventeenth century, Leibniz extended this ancient Greek idea to things that were either physical or spiritual. To Leibniz, the basis of the universe of all things—material or not—was the *monad*, a "windowless" element that has no internal structure. Leibniz considered the mutual relationships among these monads to hold the secret to all structures of creation.[11]

Cartier then invokes Bourbaki's work on the foundation of mathematics, quoting what he calls "his" (i.e., Bourbaki's) definition of set theory:

A *set* is composed of *elements* capable of having certain *properties* and having certain *relations* among themselves or with elements of other sets.

These points, or elements, are preexisting, and their combinations and mutual relations create structure in the physical universe, in Leibniz's spiritual realm, and in pure mathematics. The idea that lies at the very heart of the effort by the Greeks, followed by Descartes, Newton, Leibniz, and then by Bourbaki, is to capture the most essential properties of the physical universe and to abstract them and turn them into the basic elements of all of mathematics. It is an immensely deep and ambitious idea.

A point in Greek geometry is the intersection of two lines, and a line is the intersection of two planes. The great nineteenth-century German mathematician Bernhard Riemann, who, as mentioned earlier, created Riemannian geometry (later used by Einstein, but also extremely important in pure mathematics), proposed the idea of a surface that is stacked over a plane. Grothendieck was inspired by this idea and went a step further: he replaced the system of "open sets" used in topology—basically extensions of the idea of open intervals of numbers, such as the set of all points between 0 and 1—into something new. He defined spaces stacked over a given space.

Grothendieck then proposed an even more abstract concept: he defined a *topos* as the ultimate generalization of a space. This idea transcended the usual mathematics of the time by creating something far more abstract and general. Grothendieck allowed his topos to solve a major problem in mathematics: the nonexistence of a set containing all sets. In fact, the category of all sets constitutes a topos. In the words of Cartier, "Grothendieck claimed the right to transcribe mathematics into any topos whatever." This was the work, and a reflection of the immense confidence, of a grand master.

Grothendieck met many members of Bourbaki, and even joined the group for a time, but he worked much better in a milieu where he was the leader; so he moved to an institution the French founded almost exclusively for him: the Institut des Hautes Études Scientifiques (IHES), outside Paris. There, in the late 1950s and 1960s, he almost single-handedly redid the theory of algebraic geometry.

Descartes had first wed algebra with geometry in the seventeenth century, showing how equations could be precisely associated with geometric figures using his Cartesian coordinate system. In showing that each equation really describes a trajectory in space, Descartes laid the foundation for the field of algebraic geometry. Grothendieck extended this nascent field much further than anyone could imagine by generalizing the idea of an algebraic variety—i.e., the set of solutions to a system of equations—into something he called a *scheme*, based on an idea given to him by Pierre Cartier. The theory of schemes developed by Grothendieck weds arithmetic with geometry, realizing the century-old dream of such a unification by Cantor's nemesis, Leopold Kronecker, who once suggested this union might someday be possible.

Grothendieck is a man who thinks in great generalities and doesn't seem to care at all about details. During a lecture he once gave, in which he was making an argument based in part on prime numbers, a listener raised his hand to ask him a question.

"Can you give us a specific example?" he asked.

"An example of what?" asked Grothendieck. "You mean give an actual prime number?"

"Yes," replied the questioner.

Grothendieck, in a rush to continue his main argument, said, "Well, take 57."

Of course we know that 57 is not a prime number—it is the product of 19 and 3—but Grothendieck couldn't care less about details such as whether a number was prime. He was concerned with much grander things—in this case, an abstraction of the idea of all prime numbers. The number 57 is now affectionately dubbed "Grothendieck's prime."

Grothendieck's many major discoveries in mathematics earned him the prestigious Fields Medal—the mathematician's Nobel Prize—as well as many other awards and distinctions. Before long, his vision of mathematics brought him into conflict with Bourbaki, which he saw as a leaderless group that lacked a clear mission. At the IHES, he instituted a lecture series that rivaled that of Bourbaki in Paris. Grothendieck's School, as it was called, was becoming the place to be for any young mathematician seeking to make a name and exert an influence on the discipline.

Grothendieck nurtured a number of bright students and seemed to have a gift for determining which research topic was best suited for each student's personality and interests. He has been described as a sensitive and kind person whose home was always open to everyone. Having grown up in great deprivation, he was always concerned with the fate of the downtrodden, the poor, and the persecuted. According to Cartier, for much of his life Grothendieck has practiced "dietary restrictions that could be ascribed to choice, Judaism, or Buddhism."[12]

AS HIS FAME GREW, Grothendieck traveled throughout the world, lecturing about mathematics in such places as Brazil and the United States, where he toured for several years. At Rutgers University he met a student named Justine, who then accompanied him back to France and lived with him for two years. They had a son together, who is now a professor of statistics at an American university.

Like his father, Grothendieck was technically stateless. His birth certificate disappeared in Berlin during the war, and when he was being shuttled with his mother to and from various camps, they were considered "displaced persons." It is surprising that Grothendieck, the famous French mathematician, never asked for or received French citizenship. Throughout his life, he has been traveling on a UN passport. (Although not technically a French citizen, Grothendieck nonetheless officially changed the spelling of his name from Alexander to the French version, Alexandre.)

At the Bourbaki Congress in 1957, Grothendieck cryptically told Pierre Cartier that he was considering pursuing "activities other than mathematics." Cartier surmised that he was thinking of writing poetry or fiction, as his mother had done in the camps.[13] But apparently Grothendieck had other things in mind.

Nine years later, in 1968, the big change took place. At the very height of his career, upon turning forty, Grothendieck decided to abandon mathematics altogether. He became involved in the movement against the Vietnam War, and took part in antiwar demonstrations on the streets of Paris; then he traveled with Cartier to Vietnam in order to protest against the actions of the U.S. government. This was also the year of the Prague Spring, the Czech revolt against Soviet rule, and in France students and workers were mobilized to act against the establishment under the leadership of Danny Cohn-Bendit, known as Danny the Red.

Cartier claims that what triggered Grothendieck's strong political activism was a discovery he had made about the IHES. Apparently it was being funded in part by grants from the French Ministry of Defense. Throughout his life, Grothendieck has harbored strong antimilitary feelings, and the discovery of a connection between "his" institute and the military angered him no end.

In September 1970 Grothendieck participated in a riot in the French city of Nice and hit two policemen. He was arrested, but after the police found out that he was a prominent professor, he was quietly released. This did not diminish his political fervor. Grothendieck then started signing up to give public talks about mathematical topics—but when his turn to speak came, he would talk instead about social issues, the environment, and his rage against war and the military. Even people who agreed with him on these subjects were greatly irritated by such maneuvers. Thus the soft-spoken, kind, generally beloved mathematician began to lose friends and supporters.

That year, Grothendieck left the IHES in disgust after failing to convince the administration to reject all military-based funding. Through the help of Jean-Pierre Serre, a member of the Bourbaki group and one of the greatest mathematicians in France, he received an appointment to the Collège de France—perhaps the most prestigious of the French institutions of research and learning. But at the Collège de France, which hired him as a professor of mathematics, Grothendieck refused to lecture on the subject and, instead, talked only about political, social, and environmental issues, including his rage against nuclear arms. When his appointment expired in 1973, it was not renewed.

Grothendieck then founded a survivalist group called Survivre et Vivre (Survive and Live). He retreated into the Pyrenees, the mountainous region closest to the location of the first camp in which he was interned, Rieucros. Perhaps he had a romantic memory of the Pyrenees, seeing them from the infested, hot, flat plain of the wretched camp; perhaps he had fantasized about escaping to these mountains.

Grothendieck spent time on and off with his group in their new compound in the Pyrenees, but he also maintained a house in the town of Villecun, where he lived from 1973 to 1979. In 1977, because he participated in an incident involving people the French government called "hippies," he was accused and put on trial based on an outdated wartime French law against "consorting with foreigners." He was given a suspended sentence of six months' imprisonment and forced to pay a steep fine. This incident infuriated him even further.

Grothendieck then took up an academic position at his alma mater, the University of Montpellier—not too far from the Pyrenees—but he didn't lecture much. More and more often, he retreated alone to the mountains, and, over months of absences, his mail would pile up at the mathematics department of the university.

By the mid-1980s very few people even saw Grothendieck, as he would spend much of his time hiding in the Pyrenees. Then, in August 1991, he suddenly burned twenty-five thousand pages of his mathematical manuscripts at the University of Montpellier and abruptly left again for the mountains. Eventually he gave up his position at the university and set up a post-office box. After 1993 even the post-office box was discontinued, and Grothendieck no longer has a listed address. He seems to have disappeared without a trace.

Around the turn of the twenty-first century, two youngish researchers from a mathematics institute in Paris decided to search for the reclusive mathematician. They used clever tricks to locate a small town in the Pyrenees that they suspected might be close to Grothendieck's hideout. Then they laid a trap for the aging mathematician on a day they thought he might have to visit a location in the town in order to buy food. They succeeded, and when they encountered him, he agreed to talk and didn't run away. They made pleasant conversation for a while and even informed Grothendieck that a conjecture he had once made had just been proved by another mathematician. Very few people, if any, have seen him since. According to Cartier:

> If I can believe his most recent visitors, he is obsessed with the Devil, whom he sees at work everywhere in the world, destroying the divine harmony and replacing 300,000 km/sec by 299,887 km/sec as the speed of light![14]

Grothendieck's thoughts and writings have become increasingly bizarre. In a meditation called "The Key to Dreams," the reclusive mathematician talks about people he calls the mutants. These "mutants," whom Grothendieck describes over hundreds of pages, comprise eighteen individuals—some well known, others not—including Charles Darwin, Sigmund Freud, Walt Whitman, Mahatma Gandhi, Pierre Teilhard de Chardin, and the mathematician Bernhard Riemann. Grothendieck believes that the mutants are exceptional people who represent humanity's potential rise above the general moral decline he sees in society. He views himself as their disciple.

I spent part of the summer of 2005 trying to find Grothendieck, using clues given by the pair of French mathematicians who had found him—but to no avail. Grothendieck must be hiding very well. When I contacted Grothendieck's relatives, I was told, "[Grothendieck] is probably dead."[15] But at this writing, the great mathematician is alive.

NOTES

CHAPTER 1

1. Heath, *A History of Greek Mathematics,* 128; Plutarch, *Lives of the Noble Grecians and Romans,* chap. 3 ("Solon").
2. Ibid.
3. Heath, 129.
4. Ibid.
5. Heath, 137.
6. Ibid., 137–38.
7. Heath, 4.
8. Ibid.
9. Ibid., 5.
10. Boyer, *A History of Mathematics,* 48.
11. Ibid.
12. Ibid.
13. Heath, 77.
14. Heath, 75.

CHAPTER 2

1. Boyer, *A History of Mathematics,* 72.
2. This is proved by contradiction. Assume that the square root of 2 is rational and equal to a/b, where a and b are integers and the fraction is in lowest terms (it can't be simplified). Then $2 = a^2/b^2$ and therefore $a^2 = 2b^2$. This means that a^2 is an even number, and hence a must also be even (when you square an odd number you get an odd number). If a is even, then you can write it as $a = 2k$, where k is some integer, and we get $2 = a^2/b^2 = (2k)^2/b^2 = 4k^2/b^2$ and thus $b^2 = 2k^2$, which means that b^2 is even and hence b is even. But if both a and b are even, then the fraction a/b could not have been in lowest terms and this is a contradiction, establishing the claim that the square root of 2 is irrational.
3. Boyer, *A History of Mathematics,* 71.
4. Bell, *Men of Mathematics,* 29.
5. Ibid.

CHAPTER 3

1. Boyer, *A History of Mathematics*, 178.

CHAPTER 4

1. Boyer, *A History of Mathematics*, 210.
2. Plofker, *Mathematics in India*, 319.
3. Ibid., 73.
4. Boyer, 238.
5. Ibid., 242.
6. Escofier, *Galois Theory*, 14.

CHAPTER 5

1. Boyer, *A History of Mathematics,* 199.

CHAPTER 6

1. Boyer, *A History of Mathematics*, 256.
2. In our modern notation, for the cubic equation $x^3 + px + q = 0$, a solution is:

$$x = \sqrt[3]{-\frac{q}{2}+\sqrt{\frac{q^2}{4}+\frac{p^3}{27}}}+\sqrt[3]{-\frac{q}{2}-\sqrt{\frac{q^2}{4}+\frac{p^3}{27}}}$$

3. Ibid., 283.
4. Escofier, *Galois Theory*, 15.
5. Ibid.

CHAPTER 7

1. Boyer, *A History of Mathematics,* 303.
2. Ibid., 312.
3. Ibid., 329.

CHAPTER 8

1. Baillet, *Vie de Monsieur Descartes*, 69.
2. Bell, *Men of Mathematics*, 38.
3. Mahoney, *The Mathematical Career of Pierre de Fermat*, 26.

CHAPTER 9

1. Robert, *Leibniz, vie et oeuvre*, 23.
2. Ibid., 11.

3. Ibid., 14.
4. Belaval, *Leibniz critique de Descartes*, 67.
5. Robert, 29.
6. Bell, *Men of Mathematics*, 120.
7. Robert, 71.
8. Ibid., 31.
9. Ibid.
10. Belaval, 81.
11. Bell, 90.
12. Ibid., 97.
13. Ibid., 105.
14. Ibid., 108.
15. Ibid.

CHAPTER 10

1. Boyer, *A History of Mathematics*, 415–416.
2. Ibid.
3. Bell, *Men of Mathematics*, 139.
4. Ibid., 143.
5. Ibid., 145–146.
6. Ibid.
7. Ibid., 149.
8. Ibid.
9. Ibid., 147.
10. Ibid., 152.
11. Ibid., 221.
12. Ibid., 222.
13. Ibid., 223.
14. Ibid., 228–229.
15. Ibid., 229–230.
16. Ibid., 231.
17. In the early part of the twentieth century, Albert Einstein found non-Euclidean geometries crucial in developing his general theory of relativity.
18. Bell, 232.
19. Ibid., 234.
20. Ibid., 243.

CHAPTER 11

1. Boyer, *A History of Mathematics*, 471.
2. Bell, *Men of Mathematics*, 153.
3. Ibid.

4. Ibid., 154.
5. Ibid.
6. Ibid., 155.
7. Boyer, 470.
8. Bell, 166.
9. Ibid.
10. Ibid., 172.
11. Ibid., 173.
12. Ibid., 174.
13. Ibid., 182.
14. Boyer, 495.

CHAPTER 12

1. Escofier, *Galois Theory*, 220.
2. Ibid., 221.
3. Ibid., 222.
4. Ibid.
5. Boyer, *A History of Mathematics*, 581.
6. Ibid., 592.
7. Reprinted in Baez, "The Octonions," 145.
8. Ibid.

CHAPTER 13

1. Dauben, *Georg Cantor*, 275–276.

CHAPTER 14

1. Kanigel, *The Man Who Knew Infinity*, 168.
2. Neville, "*Srinivasa Ramanujan,*" 292–295.
3. Bell, *Men of Mathematics*, 526.

CHAPTER 15

1. Weil, *The Apprenticeship of a Mathematician*, 97.
2. Weil's work forms the mathematical appendix to Claude Lévi-Strauss's book *The Elementary Structures of Kinship* (Boston: Beacon Press, 1971), 221–233.
3. Boas, "Bourbaki and Me."
4. Personal communication from Sylvie Weil, André Weil's daughter, 2010.
5. Personal communication from Pierre Cartier, 2006.
6. Cartier, "A Mad Day's Work," 389–408.

7. Scharlau, "Who Is Alexander Grothendieck?" 93–94.

8. Jackson, "Comme Appelé du Néant—As If Summoned from the Void," 1038–1056; 1196–1212.

9. Ibid.,1039.

10. Cartier, 391.

11. Ibid., 393.

12. Ibid.

13. Ibid., 392.

14. Ibid., 393.

15. Interviews with Grothendieck's relatives, 2005.

BIBLIOGRAPHY

Aczel, Amir D. *The Artist and the Mathematician: The Story of Nicolas Bourbaki, the Mathematician Who Never Existed.* New York: Avalon, 2006.

_____. *Descartes's Secret Notebook.* New York: Broadway Books, 2005.

_____. *Fermat's Last Theorem.* New York: Basic Books, 1996.

_____. *God's Equation.* New York: Basic Books, 1999.

_____. *The Mystery of the Aleph: Mathematics, Kabbalah, and the Search for Infinity.* New York: Atria, 2000.

al-Khalili, Jim. *The House of Wisdom: How Arabic Science Saved Ancient Knowledge and Gave Us the Renaissance.* New York: Penguin Press, 2011.

al-Khowarizmi, Muhammad ibn Musa. *Robert of Chester's Latin Translation of the Algebra of al-Khowarizmi (1915).* Translated and with an introduction by Louis Charles Karpinski. Facsimile edition. Whitefish, MT: Kessinger Publishing, 2010.

Archimedes. *The Works of Archimedes.* Translated and edited by Sir Thomas Heath. Mineola, NY: Dover Publications, 2002.

Baez, John C. "The Octonions." *Bulletin of the American Mathematical Society* 39, no. 2 (2002): 145–205.

Baillet, Adrien. *Vie de Monsieur Descartes.* Paris: La Table Ronde, 2002. Reprint of the 1692 edition.

Balibar, Françoise. *Galilée, Newton, lus par Einstein.* Paris: Presses Universitaires de France, 1981.

Barrow, John D. *Pi in the Sky: Counting, Thinking, and Being.* New York: Oxford University Press, 1992.

Beeckman, Isaac. *Journal 1604–1634.* Introduction and notes by C. de Waard. 4 vols. The Hague: M. Nijhoff, 1939–1953.

Belaval, Y. *Leibniz critique de Descartes.* Paris: Editions Gallimard, 1960.

Bell, Eric Temple. *Men of Mathematics: The Lives and Achievements of the Great Mathematicians from Zeno to Poincaré.* New York: Simon and Schuster, 1986.

Beman, W. W., and D. E. Smith. *History of Mathematics.* Charleston, SC: Bibliobazaar, 2009.

Benacerraf, P., and H. Putnam, eds. *Philosophy of Mathematics.* Englewood Cliffs, NJ: Prentice-Hall, 1964.

Berlinski, David. *A Tour of the Calculus.* New York: Pantheon, 1995.

Boas, Ralph P. "Bourbaki and Me." *The Mathematical Intelligencer* 8, no. 4 (1996): 21–25.

Bolzano, Bernard. *Paradoxes of the Infinite*. New Haven, CT: Yale University Press, 1950.

Bonola, Roberto. *Non-Euclidean Geometry*. Includes original papers by J. Bolyai and N. Lobachevsky. Mineola, NY: Dover Publications, 1914.

Boria, Vittorio. "Marin Mersenne: Educator of Scientists." Ph.D. diss., American University, 1989.

Bourbaki, Nicolas. *Théorie des ensembles*. Paris: Hermann Editions, 1938.

Boyer, Carl B., and Uta C. Merzbach. *A History of Mathematics*. 2nd ed. Hoboken, NJ: Wiley, 1991.

Burton, David M. *The History of Mathematics: An Introduction*. New York: McGraw-Hill, 2010.

Cajori, Florian. *A History of Mathematical Notations*. 2 vols. Mineola, NY: Dover Publications, 1993.

Cantor, Georg. *Contributions to the Founding of the Theory of Transfinite Numbers*. Translated by Philip E. B. Jourdain. LaSalle, IL: Open Court, 1952.

Cardano, Girolamo. *The Rules of Algebra [Ars Magna]*. Translated by T. Richard Witmer. Mineola, NY: Dover Publications, 2007.

Carr, Herbert W. *Leibniz*. Mineola, NY: Dover Publications, 1960.

Cartier, Pierre. "A Mad Day's Work: From Grothendieck to Connes and Kontsevich—The Evolution of Concepts of Space and Symmetry." *Bulletin of the American Mathematical Society* 38, no. 4 (October 2001): 389–408.

Charraud, Nathalie. *Infini et Inconscient: Essai sur Georg Cantor*. Paris: Anthropos, 1994.

Chouchan, Michele. *Nicolas Bourbaki: Faits et legends*. Paris: Editions du Choix, 1995.

Christianson, Gale E. *Isaac Newton*. New York: Oxford University Press, 2005.

Cohen, Morris R., and I. E. Drabkin. *A Source Book in Greek Science*. Cambridge, MA: Harvard University Press, 1948.

Costabel, Pierre. *Démarches originales de Descartes savant*. Paris: Librairie Philosophique J. Vrin, 1982.

Cottingham, John. *Descartes*. New York: Oxford University Press, 1998.

Courant, Richard, and Herbert Robbins. *What Is Mathematics? An Elementary Approach to Ideas and Methods*. Revised by Ian Stewart. Oxford: Oxford University Press, 1996.

Couturat, Louis. *La logique de Leibniz*. New York: Georg Olms Verlag, 1985.

Croll, Oswald. *Basilica chymica*. Frankfurt: G. Tampach, 1620.

Dauben, Joseph Warren. *Georg Cantor: His Mathematics and Philosophy of the Infinite*. Princeton, NJ: Princeton University Press, 1990.

Dawson, John W., Jr. *Logical Dilemmas: The Life and Work of Kurt Gödel*. Natick, MA: A K Peters, 1997.

Deakin, Michael A. B. *Hypatia of Alexandria: Mathematician and Martyr*. Amherst, NY: Prometheus Books, 2007.

de Morgan, Augustus. *Budget of Paradoxes*. New York: Cosimo Classics (2007).

Descartes, René. *Correspondance avec Elizabeth et autres lettres*. Edited by J. M. Beyssade and M. Beyssade. Paris: Flammarion, 1989.

———. *Discours de la méthode*. Edited by, and with an introduction and notes by, Étienne Gilson. Includes the appendix, *La géométrie*. 1925. Reprint, Paris: Librairie Philosophique J. Vrin, 1999.

———. *Oeuvres philosophiques*. Vol. 1, 1618–1637. Paris: Garnier Frères, 1997.

Dolnick, Edward. *The Clockwork Universe: Isaac Newton, the Royal Society, and the Birth of the Modern World*. New York: Harper / HarperCollins Publishers, 2011.

Dunham, William W. *Journey Through Genius: The Great Theorems of Mathematics*. New York: Penguin Books, 1991.

Escofier, Jean-Pierre. *Galois Theory*. Translated by Leila Schneps. New York: Springer Verlag, 2001.

Euclid. *The Thirteen Books of the Elements*. Vol. 3, Books 10–13. Translated and with an introduction and commentary by Sir Thomas Heath. Mineola, NY: Dover Publications, 1956.

Eves, Howard. *An Introduction to the History of Mathematics*. Florence, KY: Brooks/Cole, 1990.

Faulhaber, Johann. *Arithmetischer Cubicossicher Lustgarten*. Tübingen, Germany: E. Cellius, 1604.

Fichant, Michel. *Science et métaphysique dans Descartes et Leibniz*. Paris: Presses Universitaires de France, 1998.

Field, J. V. *The Invention of Infinity: Mathematics and Art in the Renaissance*. New York: Oxford University Press, 1997.

Fourier, Joseph. *The Analytical Theory of Heat*. Mineola, NY: Dover Publications, 2003.

Fraenkel, Abraham. *Set Theory and Logic*. Boston: Addison-Wesley, 1966.

Frank, Philipp. *Einstein: His Life and Times*. New York: Knopf, 1957.

Galilei, Galileo. *Dialogue Concerning the Two Chief World Systems*. Translated by Stillman Drake. New York: Modern Library, 2001.

Gaukroger, Stephen. *Descartes: An Intellectual Biography*. Oxford: Clarendon Press, 1995.

Gilder, Joshua, and Anne-Lee Gilder. *Heavenly Intrigue*. New York: Doubleday, 2004.

Gillispie, Charles Coulston. *Pierre-Simon Laplace, 1749–1827*. Princeton, NJ: Princeton University Press, 2000.

Gleick, James. *Isaac Newton*. New York: Vintage, 2004.

Gödel, Kurt. *The Consistency of the Axiom of Choice and the Generalized Continuum-Hypothesis with the Axioms of Set Theory*. Princeton, NJ: Princeton University Press, 1940.

———. *On Formally Undecidable Propositions of Principia Mathematica and Related Systems*. Mineola, NY: Dover Publications, 1962.

Grattan-Guinness, Ivor. *The Norton History of the Mathematical Sciences.* New York: W. W. Norton, 1997.

Grothendieck, Alexandre. *Recoltes et Semailles* [Harvesting and Sowing]. Unpublished manuscript, 1986.

Grothendieck, Alexandre, and Jean-Pierre Serre. *Correspondance Grothendieck-Serre.* Providence, RI: American Mathematical Society, 2003.

Hadlock, Charles R. *Field Theory and Its Classical Problems.* Washington, DC: Mathematical Association of America, 1978.

Hallett, Michael. *Cantorian Set Theory and Limitation of Size.* New York: Oxford University Press, 1984.

Hardy, G. H. *A Mathematician's Apology.* New York: Cambridge University Press, 1992.

Hawlitschek, Kurt. *Johann Faulhaber 1580–1635: Eine Blutezeit der mathematischen Wissenschaften in Ulm.* Ulm, Germany: Stadtbibliothek Ulm, 1995.

Heath, Sir Thomas. *A History of Greek Mathematics.* Volume 1, *From Thales to Euclid.* Mineola, NY: Dover Publications, 1981.

Heilbron, John L. *Galileo.* New York: Oxford University Press, 2010.

Herivel, John. *Joseph Fourier.* Oxford: Oxford University Press, 1975.

Hobson, Ernest William. *Squaring the Circle.* Cambridge: Cambridge University Press, 1913.

Hoffman, Paul. *Archimedes' Revenge: The Joys and Perils of Mathematics.* New York: W. W. Norton, 1988.

Holton, Gerald. *Thematic Origins of Scientific Thought.* Cambridge, MA: Harvard University Press, 1973.

Ibn Khaldun. *The Muqaddimah: An Introduction to History.* Translated by Franz Rosenthal. Princeton, NJ: Princeton University Press, 1989.

Ifrah, Georges. *The Universal History of Numbers.* New York: Wiley, 2000.

Infeld, Leopold. *Whom the Gods Love: The Story of Evariste Galois.* New York: Whittlesy House, 1948.

Jackson, Allyn. "Comme Appelé du Néant—As if Summoned from the Void: The Life of Alexandre Grothendieck." Pts. 1 and 2. *Notices of the American Mathematical Society* 51, no. 9 (October 2004): 1038–1056; 51, no. 10 (November 2004): 1196–1212.

Kanigel, Robert. *The Man Who Knew Infinity: A Life of the Genius Ramanujan.* 5th ed. New York: Washington Square Press, 1991.

Katz, Victor J. *A History of Mathematics.* Boston: Addison-Wesley, 2008.

Kepler, Johannes. *Harmonices mundi.* Linz, Austria: G. Tampachius, 1619.

_____. *Mysterium Cosmographicum.* Norwalk, CT: Abaris Books, 1999.

Kline, Morris. *Mathematical Thought from Ancient to Modern Times.* New York: Oxford University Press, 1972.

Kosmann-Schwarzbach, Yvette. *The Noether Theorems: Invariance and Conservation Laws in the Twentieth Century.* Translated by Bertram E. Schwarzbach. New York: Springer Verlag, 2011.

Laplace, Pierre-Simon. *L'exposition du systeme du monde*. Paris: Librairie Arthème Fayard, 1984.

Lavine, Shaugham. *Understanding the Infinite*. Cambridge, MA: Harvard University Press, 1994.

Legendre, Adrien-Marie, and Charles Scott Venable. *Elements of Geometry After Legendre*. Whitefish, MT: Kessinger Publishing, 2008.

Leibniz, Gottfried Wilhelm. *Stamliche Schriften und Briefe*. Berlin: Akademie Verlag, 1923.

Lenoble, Robert. *Mersenne ou la naissance du mécanisme*. Paris: Librairie Philosophique J. Vrin, 1943.

Lévi-Strauss, Claude. *The Elementary Structures of Kinship*. Mathematical appendix by André Weil. Boston: Beacon Press, 1971.

Levy, Silvio, ed. *Flavors of Geometry*. New York: Cambridge University Press, 1997.

Linton, Christopher M. *From Eudoxus to Einstein: A History of Mathematical Astronomy*. New York: Cambridge University Press, 2008.

Longfellow, Ki. *Flow Down Like Silver: Hypatia of Alexandria*. Belvedere, CA: Eio Books, 2009.

Mahoney, Michael Sean. *The Mathematical Career of Pierre de Fermat*. 2nd ed. Princeton, NJ: Princeton University Press, 1994.

Mancosu, Paolo, ed. *From Brouwer to Hilbert: The Debate on the Foundations of Mathematics in the 1920s*. New York: Oxford University Press, 1997.

Mashaal, Pierre. "Le vrai General Bourbaki (1816–1897)." *Pour la Science* 2 (February–May 2000): 17.

McLeish, John. *The Story of Numbers: How Mathematics Has Shaped Civilization*. New York: Fawcett Books, 1991.

Mehl, Edouard. *Descartes en Allemagne*. Strasbourg: Presses Universitaires de Strasbourg, 2001.

Mersenne, Marin. *Quaestiones celeberrimae in Genesim*. Paris: S. Cramoisy, 1623.

Meserve, Bruce E. *Fundamental Concepts of Geometry*. Mineola, NY: Dover Publications, 1983.

Monge, Gaspard. *Géometrie descriptive*. Paris: J. Klostermann, 1811.

Moore, Gregory H. *Zermelo's Axiom of Choice: Its Origins, Development, and Influence*. New York: Springer Verlag, 1982.

Neuenschwander, Dwight E. *Emmy Noether's Wonderful Theorem*. Baltimore, MD: Johns Hopkins University Press, 2010.

Neville, Eric Harold. "Srinivasa Ramanujan." *Nature* 149 (March 14, 1942): 292–295.

Newton, Isaac. *The Principia: Mathematical Principles of Natural Philosophy*. Berkeley, CA: University of California Press, 1999.

Nicastro, Nicholas. *Circumference: Eratosthenes and the Ancient Quest to Measure the Globe*. New York: St. Martin's Press, 2008.

Pickover, Clifford A. *The Math Book: From Pythagoras to the 57th Dimension, 250 Milestones in the History of Mathematics.* New York: Sterling Publishing, 2009.

Plofker, Kim. *Mathematics in India.* Princeton, NJ: Princeton University Press, 2009.

Plutarch. *Lives of the Noble Grecians and Romans.* Edited by Arthur Clough. 1859. Reprint, Oxford: Benediction Classics, 2010.

Poincaré, Henri. *Science and Method.* Toronto: University of Toronto Libraries, 2011.

_____. *The Value of Science: Essential Writings of Henri Poincaré.* Edited by Stephen Jay Gould. New York: Modern Library, 2001.

Quine, Willard Van Orman. *Set Theory and Its Logic.* Cambridge, MA: Harvard University Press, 1963.

Ramsey, Frank P. *The Foundations of Mathematics and Other Logical Essays.* Edited by R. Braithwaite. London: Kegan Paul, 1931.

Reichenbach, Hans. *The Philosophy of Space and Time.* Mineola, NY: Dover Publications, 1958.

Robert, Jean-Michel. *Leibniz, vie et oeuvre.* Paris: Univers Poche, 2003.

Roth, Peter. *Arithmetica philosophica.* Nuremberg, Germany: J. Lantzenberger, 1608.

Russell, Bertrand. *The Philosophy of Leibniz.* London: Gordon and Breach, 1908.

Scharlau, Winfried. "Who Is Alexander Grothendieck?" *Notices of the American Mathematical Society* 55, no. 8 (September 2008): 930–941.

Scheiner, Christoph. *Oculus hoc est: Fundamentum opticum.* Innsbruck, Austria: D. Agricola, 1619.

Schoenflies, Arthur. *Entwickelung der Mengenlehre.* Leipzig, Germany: B. G. Teubner, 1913.

Schrecker, Paul, ed. G. W. *Leibniz: Opuscules philosophiques choisis.* Paris: Librairie Philosophique J. Vrin, 2001.

Schwartz, Laurent. *Un mathématicien aux prises avec le siècle.* Paris: Odile Jacob, 1997.

Scribano, Emanuela. *Guida alla lettura della Meditazioni metafisiche di Descartes.* Rome: Editori Laterza, 1997.

Senechal, Marjorie. "The Continuing Silence of Bourbaki: An Interview with Pierre Cartier." *The Mathematical Intelligencer* 20, no. 1 (1998): 22–28.

Shea, William R. *The Magic of Numbers and Motion: The Scientific Career of René Descartes.* Canton, MA: Science History Publications, 1991.

Sigler, Laurence E. *Fibonacci's Liber Abaci.* New York: Springer Verlag, 2003.

Simon, Gerard. *Kepler, astronome, astrologue.* Paris: Editions Gallimard, 1992.

Stanley, Thomas. *Pythagoras: His Life and Teachings.* Lake Worth, FL: Ibis Press, 2010.

Stillwell, John. *Mathematics and Its History.* 3rd ed. New York: Springer Verlag, 2010.

Struik, Dirk Jan. *A Concise History of Mathematics*. 4th ed. Mineola, NY: Dover Publications, 1987.

Szpiro, George. *Kepler's Conjecture*. Hoboken, NJ: Wiley, 2003.

Tartaglia, Niccolo, and Guillaume Gosselin. *L'arithmetique de Nicolas Tartaglia, Brecian, 1613*. Whitefish, MT: Kessinger Publishing, 2010.

Van Peursen, C. A. *Leibniz*. London: Faber and Faber, 1969.

Varaut, Jean-Marc. *Descartes: Un cavalier francais*. Paris: Editions Plon, 2002.

Verbeek, Theo, ed. *René Descartes et Martin Schoock: La querelle d'Utrecht*. Paris: Les Impressions Nouvelles, 1988.

Wallace, David Foster. *Everything and More: A Compact History of Infinity*. New York: W. W. Norton, 2003.

Weil, André. *The Apprenticeship of a Mathematician*. Translated by Jennifer Gage. Boston: Birkhäuser, 2004.

Weil, Sylvie. *At Home with André and Simone Weil*. Translated by Benjamin Ivry. Evanston, IL: Northwestern University Press, 2010.

Wells, David. *The Penguin Dictionary of Curious and Interesting Numbers*. New York: Penguin Books, 1987.

Westfall, Richard S. *Never at Rest: The Biography of Isaac Newton*. Cambridge: Cambridge University Press, 1983.

Weyl, Hermann. *Philosophy of Mathematics and Natural Science*. Princeton, NJ: Princeton University Press, 2009.

Whitehead, Alfred North, and Bertrand Russell. *Principia Mathematica*. 3 vols. Cambridge: Cambridge University Press, 1910–1913.

Wolfe, Harold E. *Introduction to Non-Euclidean Geometry*. New York: Holt, Rinehart and Winston, 1945.

Wootton, David. *Galileo: Watcher of the Skies*. New Haven, CT: Yale University Press, 2010.

PHOTO CREDITS

Grand; Rigaud Hyacinthe.jpg/Source: Museo del Prado/User: WGA.hu; 123: Gottfried Wilhelm Leibniz c1700.jpg/Source: Archiv der Berlin-Brandenburgischen Akademie der Wissenschaften/User: AndreasPraefcke; 128: Woolsthorpe Manor.jpg/User: Xander89; 129: Newtons room in Cambridge.jpg/Source: www.hypatiamaze.org/User: QWerk; 132: Newton—Principia (1687), title, p. 5, color.jpg/Source: Service Commun de la Documentation University of Strasbourg/User: Piero; 136: Old University Basel.jpg/User: Gulliveig; 144: Leonhard Euler 2.jpg/User: Haham hanuka; 145: Catherine II Russia-v2-front.jpg/Source: Alfred Rambau's *The History of Russia,* volume 2 (1898)/User: Moverton; 148: Bendixen—Carl Friedrich Gauß, 1828.jpg/Source: "Astronomische Nachrichten," 1828/User: Rabe!; 153 (top left): RechtwKugeldreieck.svg/User: Stannered; 153 (top right): Hyperbolic triangle.svg/User: Kieff; 153 (bottom): Argandgaussplane.gif/User: LeonardoG; 160: Andrea Appiani 002.jpg/Source: The Yorck Project: *10.000 Meisterwerke der Malerei.* DVD-ROM, 2002. Distributed by DIRECTMEDIA Publishing GmbH; 162: Académie des Sciences 1671.jpg/Source: an engraving by Sebastien Le Clerc from *Mémoires pour servir a l'Histoire Naturelle des Animause* (Paris, 1671)/User: Papa November; 168: Joseph Louis Lagrange.jpg/User: Kelson; 170: LouisXVIExecutionBig.jpg/Source: Museum of the French Revolution/User: Bkwillwm; 171: Alembert.jpg/User: Archaeodontosaurus; 172: AduC 197 Laplace (P.S., marquis de, 1749–1827).JPG/Source: Augustin Challamel, Desire Lacroix, *Album du centenaire,* Paris: Jouvet & Cie, éditeurs, 1889/User: Havang(nl); 178: Bataille des pyramides ag1.png/Source: A. Hugo, *France Militaire Histoire des armées françaises de terre et de mer de 1792 à 1833 Tome 1,* Delloye, Paris, 1835/User: Greatpatton; 180: Eugène Delacroix—La liberté guidant le peuple.jpg/User: Aavindraa; 183: Evariste galois.jpg/User: Anarkman; 189: E. Galois Letter.jpg/Source: Iyanaga, Shokichi, "ガロアの時代 ガロアの数学 第一部 時代篇," Springer-Verlag Tokyo, 1999; 193: Clifford William Kingdon desk.jpg/Source: *Lectures and Essays by the Late William Kingdon Clifford, F.R.S.,* volume 2; 195: Dunsink.jpg/User: Jaqian; 196: William Rowan Hamilton portrait oval combined.png/User: Quibik; 200: SunShiningThroughDustInWood.jpg/Author: Andreas Tille; 205 (top): Uni-Halle-1836.jpg/User: Torsten Schleese; 205 (bottom): Karl Weierstrass 3a.jpg/Source: Hermann von Helmholtz-Zentrum für Kulturtechnik (HZK)/User: Skraemer; 207: Georgcantor01.png/User: Jan Arkesteijn; 218: Burlington House in 1854.jpg/User: Merchbow; 222: Ghhardy@72.jpg/User: Macdonald-Ross; 224: Felix Christian Klein.jpg/User: Sobi3ch; 225 (left): Moebius Surface 1 Display.png/User: Inductiveload; 225 (right): 225 (right): KleinBottle-01.svg/User: AnonMoos; 226: Goettingen Audi Max 01.jpeg/Source: Städtisches Museum Göttingen (Hrsg.): *Göttingen. Das Bild der Stadt in historischen Ansichten,* S. 321, Göttingen 1996/User: Archaeotect; 227: Noether.jpg/User: Anarkman; 231: Einstein 1921 portrait2.jpg/Source: Historisches Museum Bern Einstein Museum/User: Craigboy; 234: Chapelle de St Eutrope.JPG/User: Castanet; 236: ENS dvur.JPG/

Author: Jan Sokol; 238: Bourbaki congress1938.jpg/Source: planetmath.org/ User: Andrew69; 245: Bourbaki, Theorie des ensembles maitrier.jpg/Author: Maitrier; 250: Alexander Grothendieck.jpg/Source: http://owpdb.mfo.de/ detail?photoID=1452/Upload by AEDP

Courtesy of Wikipedia: 241: Henri Poincaré by H Manuel.jpg/Source: *Scientific Identity: Portraits from the Dibner Library of the History of Science and Technology*/User: Sv1xv

INDEX

Note: Page numbers in *italics* include illustrations, photographs, and captions.